Advances in Electroceramic Materials

Advances in Electroceramic Materials

Ceramic Transactions, Volume 204

A Collection of Papers Presented at the 2008 Materials Science and Technology Conference (MS&T08)
October 5–9, 2008
Pittsburgh, Pennsylvania

Edited by
K. M. Nair
D. Suvorov
R. W. Schwartz
R. Guo

The American Ceramic Society

WILEY

A John Wiley & Sons, Inc., Publication

Published by John Wiley & Sons, Inc., Hoboken, New Jersey.
Published simultaneously in Canada.

Library of Congress Cataloging-in-Publication Data is available.

ISBN 978-0-470-40844-5

Printed in the United States of America.

10 9 8 7 6 5 4 3 2 1

Contents

Preface

New areas of materials technology development and product innovation have been extraordinary during the last few decades. Our understanding of science and technology behind the electronic materials played a major role in satisfying the social needs by developing electronic devices for automotive, telecommunications, military and medical applications. The electronic technology development still has an enormous potential role to play in developing future materials for these consumer applications. Miniaturization of electronic devices and improved system properties will continue during this century to satisfy the increased demands of our society particularly in the area of medical implant devices, telecommunications and automotive markets. Cost-effective manufacturing technology development should be the new areas of interest due to the high growth of market in countries like China and India. By working together, international scientific societies can play a major role for development of new manufacturing technology.

The materials societies understand their social responsibility. For many years, The American Ceramic Society (ACerS) has organized several international symposia covering many aspects of the advanced electronic material systems by bringing together leading researchers and practitioners of electronics industry, university and national laboratories and publishing the proceedings of the conferences in their Ceramic Transactions series.

This volume contains a collection of papers from the Advanced Dielectric Materials and Electronic Devices and Electroceramics Technologies symposia held during MS&T08—a joint meeting between ACerS, AIST, ASM International, and TMS—held at the David L. Lawrence Convention Center, Pittsburgh, Pennsylvania, USA, October 5–9, 2008.

The editors acknowledge and appreciate the contributions of the speakers, conference session chairs, manuscript reviewers and ACerS staff for making this endeavor a successful one.

K. M. Nair, E.I. duPont de Nemours & Co., Inc, USA
D. Suvorov, Jozef Stefan Institute, Solvenia
R. W. Schwartz, Missouri University of Science and Technology, USA
R. Guo, University of Texas, USA

Design, Synthesis and Characterization

CERAMIC-POLYMER DIELECTRIC COMPOSITES PRODUCED VIA DIRECTIONAL FREEZING

E.P.Gorzkowski, and M.-J. Pan
Naval Research Laboratory
4555 Overlook Ave., SW
Washington, DC 20375

ABSTRACT

The freeze casting method was successfully used to create ceramic-polymer composites with the two phases arranged in an electrically parallel configuration. The result is a novel composite that exhibits dielectric constant (K) of up to 4000 for PMN-10PT while maintaining low dielectric loss (< 0.05). The finished composites not only exhibit the high dielectric constant of ferroelectric ceramics but maintain the flexibility and ease of post-processing handling of polymer materials. Graceful failure of these samples was observed during dielectric breakdown testing as well as high d_{33} and good hysteresis behavior. In fact the PZT-5A samples had a d_{33} value of ~250 pC/N and a remnant polarization of 15 $\mu C/cm^2$.

INTRODUCTION

A recent article in Science[1] demonstrates the fabrication of nacre-like laminar ceramic body using a novel ice template process. This technique entails freezing an aqueous ceramic slurry uni-directionally along the longitudinal axis of a cylindrical mold to form ice platelets and ceramic aggregates. Given the proper conditions, which include slurry viscosity, percentage water, temperature gradient between the top and bottom of the mold, and starting temperature, the ice platelets are aligned in the temperature gradient direction. The proper starting temperature and temperature gradient must be maintained so that homogeneous freezing occurs and hexagonal ice is formed. This allows the ice front to expel the ceramic particles in such a way to form long range order for both the ceramic and the ice. Upon freeze drying, the ice platelets sublime and leave a laminar ceramic structure with long empty channels in the direction of the temperature gradient. Subsequently the green ceramic body is sintered to form the final microstructure.

This article focuses only on the mechanical properties of the ceramic body, but a ceramic-polymer composite with excellent dielectric properties may be possible by adapting the technique. The adaptation involves 1) using a high K material as the ceramic phase, 2) infiltrating the space between ceramic lamellae with a polymer material, and 3) applying electrodes perpendicular to the ceramic-polymer alignment direction to form an electrically parallel composite dielectric. In this way, the resultant material should exhibit a dielectric constant up to two orders of magnitude higher than that of existing polymer-based dielectrics.

EXPERIMENTAL PROCEDURES

Ceramic slurries were prepared by mixing purified water with 2 wt% of the ammonium polymethacrylate dispersant, Darvan C, (R.T. Vanderbilt Co., Norwalk, CT), 1 wt% of polyvinyl alcohol (Alfa Aesar, Ward Hill, MA), and 68 wt% PZT-5A ceramic (Morgan Electroceramics, Bedford, OH) or 72 wt% PMN-10PT ceramic (made via the Columbite method). Slurries were ball-milled in a high density polyethylene bottle for 12 h with zirconia milling media and de-aired in a vacuum desiccator.

Freezing of the slurries was accomplished by pouring them into a Teflon mold (1.5 in. diameter, 0.75 in. tall) and cooled using a custom built freezing setup. The mold is placed between two copper rods that are cooled by liquid nitrogen to – 60 °C at 5 °C/min. There are band heaters attached to the copper rods in order to control the cooling rate and temperature gradient between the copper rods (10 °C). The samples were freeze-dried (Freeze Dryer 2.5, Labconco, Kansas City, MO) for 24 h. Samples were then removed from the mold for annealing.

Binder burnout and bisque firing was done by heating the samples at 1.2 °C/min to 300 °C, 0.1 °C/min to 350 °C, 0.6 °C/min to 500 °C, 5 °C /min to 900 °C , and finally, a 1 h dwell at 900°C. The samples were then sintered at 1150 °C for 2 h. Each cylindrical sample was then infiltrated with Epotek 301 Epoxy (Epoxy Technology, Billerica, MA) under vacuum creating a composite that is 25 vol% ceramic 75 vol% polymer. Smaller cylindrical plate capacitor samples were cut and prepared for dielectric testing. This entailed lapping the samples using 400 and 600 grit SiC slurry to create flat parallel faces. Some samples were gold coated for capacitance measurement, while others were masked for breakdown and d_{33} measurements. The dielectric constant and loss were measured using an HP 4284A at 0.1, 1, 10, and 100 kHz from 150 down to -60 °C. The breakdown measurements were made using a Hipot tester (QuadTech) at 100 V/s. The d_{33} measurements were performed on the PZT-5A samples using a Berlincourt piezo d_{33} meter (Channel Products INC., Hesterland, OH) after being poled at 80 °C for 5 min. at 30 kV/cm.

Pieces from each of the various samples were mounted onto a stub, carbon coated and masked with conductive tape for Scanning Electron Microscopy (SEM). Images of these surfaces were obtained using a Leo 1550 SEM. Light Optical images were taken with a Nikon microscope (Nikon Instruments, Melville, NY).

RESULTS AND DISCUSSION

After freeze-drying the green samples were fragile but easily transportable. The sample shrinks approximately 2 % in all dimensions after the binder burnout and bisque firing, but the strength of the sample increases dramatically. Once sintered the samples shrink in the lateral and longitudinal directions by ~40%. This is due to the reduction of porosity in the ceramic platelets as the size of the channels between the plates does not decrease after sintering. The SEM image in Figure 1 shows that there is alignment of the PZT-5A ceramic lamellae. It is important to note that the PMN-10PT microstructure looks similar to the PZT-5A, and was not included to avoid repetition.

The microstructure of the sintered samples which consists of ceramic plates aligned in the direction of the temperature gradient. In previous studies,[2] interconnects formed between the ceramic plates, but better care was taken to make sure that the cooling rate was controlled. By controlling the cooling rate the ice front does not reach supersaturation of the ceramic and thus no particle repulsion which causes the local ice crystal front to split leaving behind an agglomerate of ceramic particles.[3] In addition the platelets are not exactly parallel to the temperature gradient. This is due to the differences between the imposed and the preferred growth directions. The preferred growth direction is controlled by the system i.e. interfacial energies while the imposed growth direction is highly dependent on the temperature gradient.[4,5] If the temperature gradient is too low then the preferred growth direction dominates and thus the platelets grow a few degrees off of the temperature gradient direction. A larger temperature gradient can correct this problem and will be used in future experiments.[3]

Figure 1 SEM image of the microstructure of freeze-cast PZT-5A sample.

In order to determine the dielectric properties of this composite the fired ceramic was infiltrated with epoxy and cut into smaller pieces perpendicular to the freezing direction. After polishing and electroding with gold the dielectric properties were measured. The dielectric constant versus temperature can be found in Figure 2 for (a.) the PZT-5A and (b.) the PMN-10PT samples. The peak value of dielectric constant for the PZT-5A was 500 and 4000 for the PMN-10PT. In both cases the dielectric constant is 2 orders of magnitude higher than conventional composite capacitors and about 50% lower than that of the sintered ceramic of the same compositions. Additionally, the dielectric loss of these samples is less than 0.05 for most frequencies which is lower than sintered ceramics but not as good as polymers. This means that the ice template method is a viable way to produce high dielectric constant composite capacitors.

The epoxy used in these experiments were not flexible enough to bend by hand, but any thermoplastic or mixable thermoset polymer can be infiltrated. Therefore, the composite can maintain the flexibility and ease of post-processing handling of polymer materials. In fact, flexible polymer/ceramic capacitors with high dielectric constant and high breakdown strength can be produced.

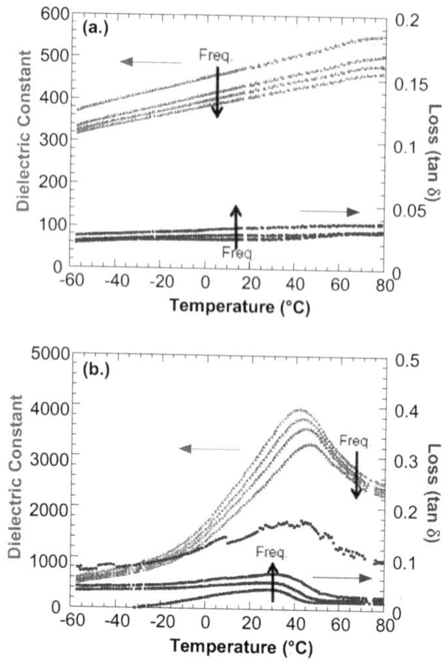

Figure 2 Dielectric constant versus temperature data for (a.) the PZT-5A and (b.) the PMN-10PT samples.

In order to be confident that the alignment of the particles caused the higher dielectric constant, a fully dense ceramic and a random composite sample was created. The ceramic sample was created from the same powder batch as the aligned composite as well as pressure less sintered at the same temperature and time. The random composite was created by mixing the same volume of ceramic powder that was in the aligned composite in epoxy and allowed to cure. Figure 3 shows the dielectric constant versus temperature of the various samples for comparison purposes. It can be seen that the ceramic value as expected is the highest and the random composite is the lowest. In fact the ceramic sample is two orders of magnitude higher than the aligned composite (freeze-cast sample) and the random composite is two orders of magnitude higher than the aligned composite. This shows that the alignment is responsible for the increase in the dielectric constant.

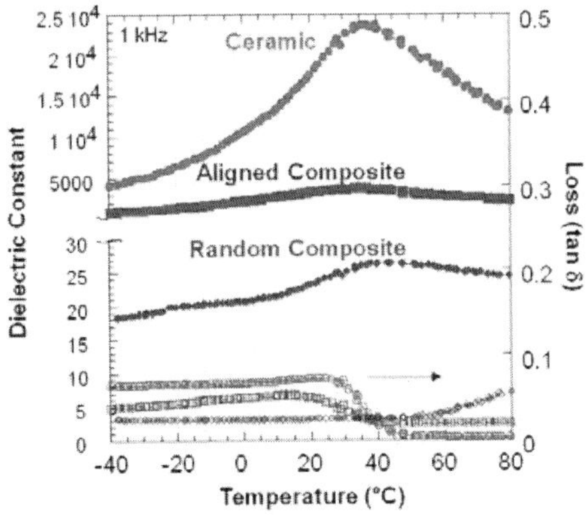

Figure 3 Dielectric Constant at 1 kHz versus temperature plot that compares PMN-10PT ceramic, freeze-cast, and random composites.

Another verification that the alignment process works well was found by observing other properties of these samples. Figure 4 shows the polarization versus field plots for the (a.) PZT-5A and the (b.) PMN-10PT samples. Both samples show a ferroelectric behavior as would be expected. The values for the coercive field, remnant polarization, and peak polarization are comparable to commercially available ceramic samples with the same respective compositions.

Breakdown measurements were also taken of these samples. These tests were done by increasing the voltage on the sample until a breakdown event occurred. The same sample was ramped up again until another breakdown event occurs and this process was repeated up to 50 times, which can be seen in Figure 5. No catastrophic failure or fail-short was observed for either composition over this testing range. Since the area around the breakdown is healed like in most polymer capacitors, voltage can be re-applied. This means that these composites fail in a graceful manner though the mechanism was not studied further. In the case of the PMN-PT/Epoxy composite the breakdown strength increased as the number of breakdown events increased. This is most likely due to the established "weakest link" theory, where breakdown occurs at the weakest point of the sample. Since the next weakest spot, the area where the next breakdown occurs is stronger than the first the breakdown voltage goes up.

The last property that was measured for the PZT-5A composites was the piezoelectric coefficient, d_{33}. In this case the value was ~250 pC/N. The ceramic value is 300-450 pC/N so the composite sample performs very well. Overall it seems that the freeze-casting method provides a viable way to make composite capacitors with excellent dielectric and piezoelectric properties. The only drawback to make this process viable for large scale manufacturing is the freeze-drying step. This may be avoided by using non-aqueous slurries. For example, camphene has been used to make slurries like this because camphene sublimes at room temperature and is liquid at 50 °C. Therefore, future studies will include the study of non-aqueous slurries.

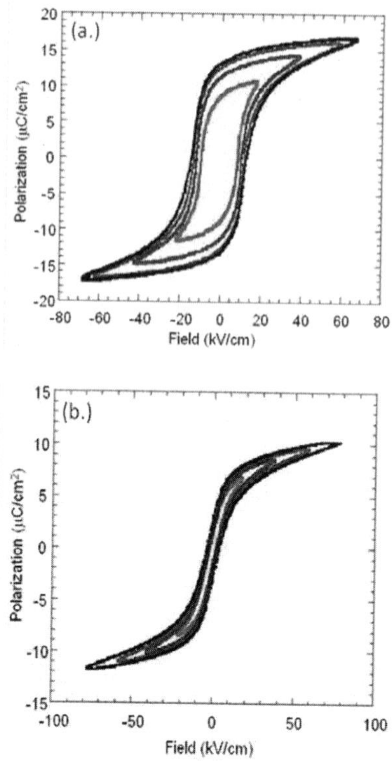

Figure 4 Polarization versus Field curves for (a.) the PZT-5A and (b.) the PMN-10PT samples.

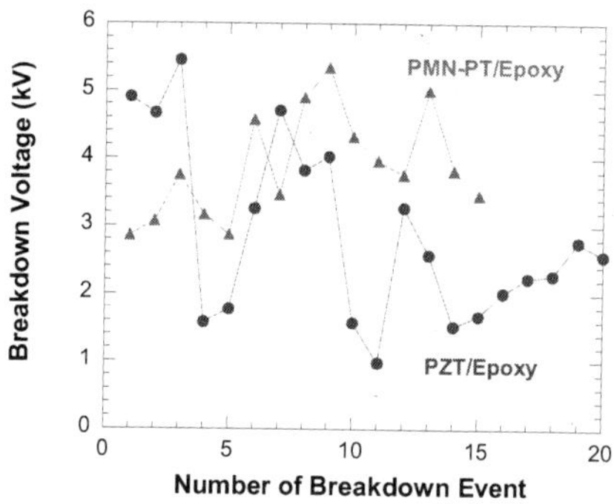

Figure 5 Breakdown voltage results for PZT-5A and PMN-10PT freeze-cast samples showing graceful failure.

CONCLUSION

The freeze casting method was successfully used to create ceramic-polymer composites with the two phases arranged in an electrically parallel configuration. The result is a novel composite that exhibits dielectric constant (K) up to two orders of magnitude higher than that of composites with ceramic particles randomly dispersed in a polymer matrix while maintaining low dielectric loss (< 0.05). The finished composites not only exhibit the high dielectric constant of ferroelectric ceramics but maintain the flexibility and ease of post-processing handling of polymer materials. Graceful failure of these samples was observed during dielectric breakdown testing as well as high d_{33} and good hysteresis behavior.

REFERENCES

[1] S. Deville, E. Saiz, R. Nalla, and A. Tomsia, "Freezing as a Path to Build Complex Composites," *Science*, **311**, 515-518 (2006).

[2] E. P. Gorzkowski and M. J. Pan, "Novel Ceramic-Polymer Composites via the Freeze Casting Method," *Proceedings of the 13th US-Japan Seminar on Dielectric and Piezoelectric Ceramics*, pp. 212-215, November 4-7, 2007.

[3] S. Deville, E. Saiz, and A. Tomsia, "Ice-templated Porous Alumina Structures," *Acta Materialia*, **55**, 1965-1974 (2007).

[4] G.W. Young, S.H. Davis, and K.J. Brattkus, "Anisotropic Interface Kinetics and Tilted Cells in Unidirectional Solidification," *J. Cryst. Growth*, **83** 560-71 (1987).

[5] K. Nagashima and Y.J. Furukawa, "Nonequilibrium effect of anisotropic interface kinetics on the directional growth of ice crystals," *J. Cryst.Growth*, **171** 577-85 (1997).

LOW-TEMPERATURE FABRICATION OF HIGHLY LOADED DIELECTRIC FILMS MADE OF CERAMIC-POLYMER COMPOSITES FOR 3D INTEGRATION

Jong-hee Kim, Eunhae Koo, Young Joon Yoon, and Hyo Tae Kim*

Fusion and Convergence Technology Division
Korea Institute of Ceramic Engineering and Technology, Seoul 153-801, Korea

ABSTRACT

Ceramic-organic composite thick films over 50 vol. % of dielectric powder loading has been fabricated at the temperatures lower than 300°C using several processing technologies. The objective of this challenge is to overcome the genetic obstacles of conventional ceramics as well as organic technologies in ceramic packaging. The ceramic technology has been confronted with severe shrinkage, brittleness and high temperature processing. The organic based, i.e. PCB technology with FR-4 epoxy also was not free from low dielectric properties and reliability issues, especially at the RF and microwave applications. In this work, ink-jet printing and aerosol deposition method were used in order to form a highly loaded powder packing in the thick film bed, and then low loss organic resins were infiltrated into the dielectric powder packing layer followed by thermal treatment under 300°C for completely hermetic monolith film. High solid loading of dielectrics with submicron size powders up to 68 vol. % was obtained by ink-jet printing via minimized polymer vehicles and controlled particle shape, size and distribution. Dielectric properties of thus obtained ceramic-organic composite film exhibited dielectric constants of 4.0 – 4.6 and Q factors of 248±34 at 1MHz.

INTRODUCTION

Forming thick film dielectric layer is a key process in the fabrication of integrated passive components and multilayer package modules. Most generally known conventional thick film technology is a tape casting method, and a low-temperature co-fired ceramic (LTCC) technology which is widely adopted in the RF and microwave device and packages. LTCC technology could deliver many advantages; i) low-temperature sintering, thus energy saving process, ii) high performance due to using highly conductive electrode metals a such as Ag or Cu, iii) rapid prototyping, and (iv) integration of passives. Regardless of those benefits, still LTCC based devices and packages are used in the limited applications due to cost and process compatibility compared with the widely used PCB technologies. Though sintering temperature of LTCC was lowered to 900°C, the genetic processing problems existed in the conventional ceramic technology such as severe shrinkage, brittleness and co-firing issues like inter-diffusion between dissimilar materials still remain further solutions.

In this work, we attempted to combine the advantages of both ceramic and polymer based dielectric materials into a ceramic/polymer composite structure by establishing backbones of highly loaded ceramic fillers and infiltrating low-loss polymers at the temperature under 300°C. Ink-jet printing [1] and aerosol deposition method [2] were used in order to form a highly loaded powder packing in the thick film bed, and then low loss organic resins were infiltrated into the dielectric powder packing layer followed by thermal treatment for completely hermetic monolith film.

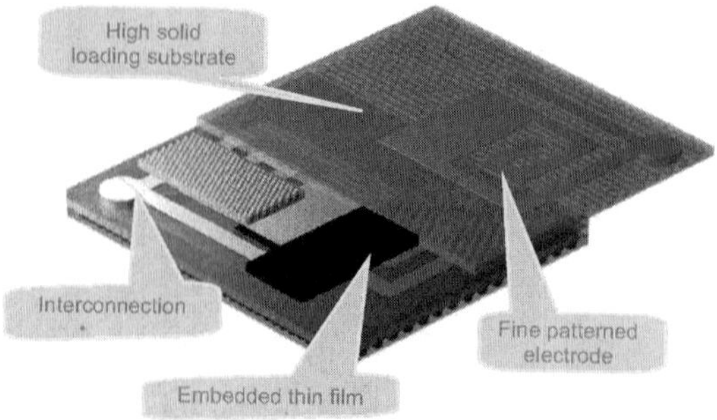

Figure 1 Schematic diagram of 3D integrated ceramic–organic hybrid module.

The schematic 3D integrated module by ceramic-organic composite layers consisted of dielectric insulating layer as a basic substrate, micro-patterned conducting circuits, embedded passives, and interconnects such as via holes as shown in Figure 1. In this paper, we only describe the formation of dielectric thick film layers with different approaches, and among them inkjet printed technology will be emphasized such that the proposed approach will meet the PCB compatible process.

EXPERIMENTAL PROCEDURE

For better dielectric properties, high solid loading dielectric layers are required in the ceramic-polymer composite structure. Selection of controlled particle size and distribution of dielectric powders is the first requirement for highly dense powder packing, and then selection of proper polymers with good dielectric properties and rheological characteristics is also an important factor for successful infiltration process. As a basic dielectric material, aluminum oxide was chosen first because of the compositional simplicity and relatively decent dielectric properties as a substrate material [3]. Aluminum oxide is also a very convenient material due to diverse powder manufacturing sources and particle morphologies in the commercial bases. Thick film forming technology we have chosen are inkjet printing and aerosol deposition method (ADM). Figure 2 illustrates the high density powder loading scheme; i) simulation of high density packing model, ii) synthesis or preparation of spherical powders, iii) powder deposition process, iv) adding or co-using of precursors for further increase in solid loading, and v) resin infiltration for pore filling to get fully hermetic layer structure.

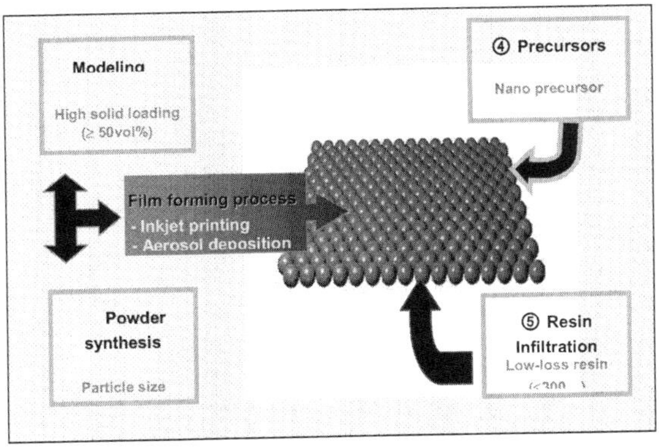

Figure 2 Experimental procedure of low- temperature film forming process.

It has been demonstrated that maximum particle packing density can be reached up to 95% in the modeling of a high density particle packing using mixed spheres of different sizes (size ration 320/39/7/1), though they used much larger sizes ranging from 0.61 to 6.07mm for experimental convenience [4]. However, there are many factors which hinder a high density in reality of thick film forming process such as dispersion, particle sedimentation, binders, solvents and other additives. In this experiment, we used two kinds of alumina powders due to the availability of commercial powders, non-spherical with nearly mono sized and spherical with multi-modal distribution (Figure 3).

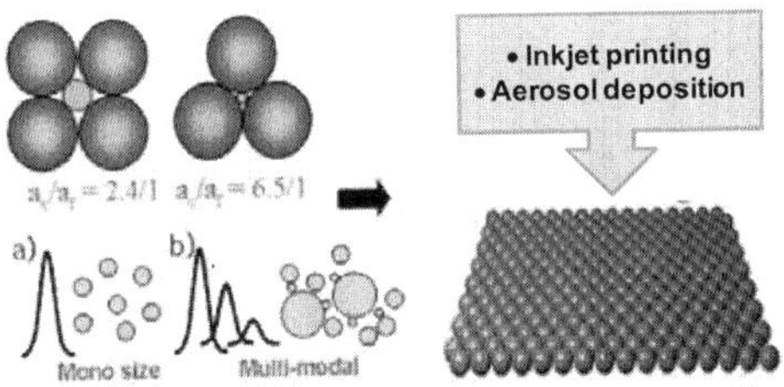

Figure 3 Selection of powders particle size and distribution for dense solid packing.

RESULTS AND DISCUSSION

Alumina/Polymer resin composite thick films by Inkjet printing

Thick film formation by inkjet printing generally requires several procedures; substrate preparation, cleaning, ink formulation, jetting, drying, resin infiltration, and curing. The recommended rheological properties of ink suspension are; i) surface tension 25-45 mN/m, ii) viscosity ranging 3-20 mPa·s, iii)

Newtonian flow behavior. Alumina inks were prepared using mixed solvent of 75 vol% water and 25 vol% formamide (FA) as a drying control agent. The sizes of alumina powders are 0.2μm (ASFP-20, Denka), 0.3μm (AKP-30, Sumitomo) respectively. The details of ink formulation and fabrication method are described in elsewhere [5]. Printing of alumina ink was carried out using UJ 200 (manufactured by Unijet Inc.) with 50μm orifice diameter of a piezoelectric nozzle fabricated by Microfab Technologies, Inc. The printer head was mounted onto a computer-controlled three-axis gantry system capable of the movement accuracy of ±5μm. A CCD camera with strobe LED light was used to characterize the size and shape of individual droplet. The volume of expelled droplets was 150-160pL, traveling with the velocity of 2.5-3.2m/s. The microstructure of alumina dielectrics on Si wafer was characterized using a field emission scanning electron microscope (model: JSM6700F, JEOL). The electrical properties were measured using impedance analyzer (model: HP4194A, Agilent Technologies Inc.).

The ceramic ink formulation contains 10 wt% alumina powders loading and added a 10 vol% of dispersant (BYK-111, BYK Chemicals) to the alumina amount. Figure 4 shows the jetting patterns of alumina ink droplets according to the desired patterns. The diameter of jetting droplets was 61-65μm, while the printed droplet size was about 130um after drying. A jetting mode in Figure 4 (c) was chosen for low-K dielectric layer forming. The printing layer thickness of one step jetting was about 1.3um and the total layer thickness slightly decreased with multiple jetting, i.e. 5.12um at 5 time jetting and 8.03um at 10 time jetting. The polymer resins used are epoxy resin, benzocyclobutene (BCB), polyphenilene oxide resin (PPO), respectively. The BCB resin was dissolved in mesitylene solution and PPO resin was also dissolved in the toluene solution for 10 wt% solid resin ink. The epoxy resin solution, however, was fabricated in-house by using a blending of noblack and bromme containing epoxies (YDB-400, YD-128, KBPN-120 with ratio of 60:20:20 wt%, Kukdo Chemicals) and bis-phenol A (KBH-L2121) hardening agent and hardening catalyst (2MI) with the ratio of 58: 40: 2 wt%. The curing temperatures of BCB, epoxy, PPO resins were 150°C, 270°C, and 300°C, respectively.

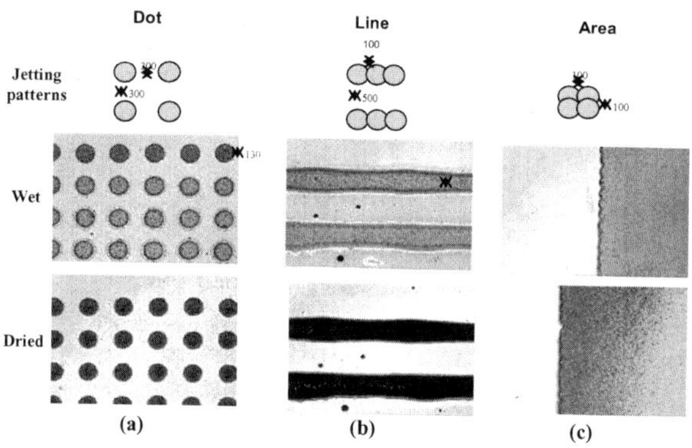

Figure 4 Typical jetting patterns of alumina inks printed on silicon substrate.

The microstructures of inkjet printed alumina films with different particle size and shapes presented in Figure 5 shows that the alumina powders with multi-modal and spherical shape provide smoother film surface and denser particle packing. Table 1 summarized the evaluation of powder packing density in the film and about 68 vol% packing density was obtained by using powders with multi-modal and spherical shape. Figure 6 shows the microstructure of alumina thick films before and after resin (BCB) infiltration demonstrates that the polymer resin was fully infiltrated into the alumina powder bed layer. The quality factor of the alumina films with PPO resin measured by impedance analyzer was about 248 at 1 MHz which shows a potential usage as a dielectric film for RF device applications (Table 2).

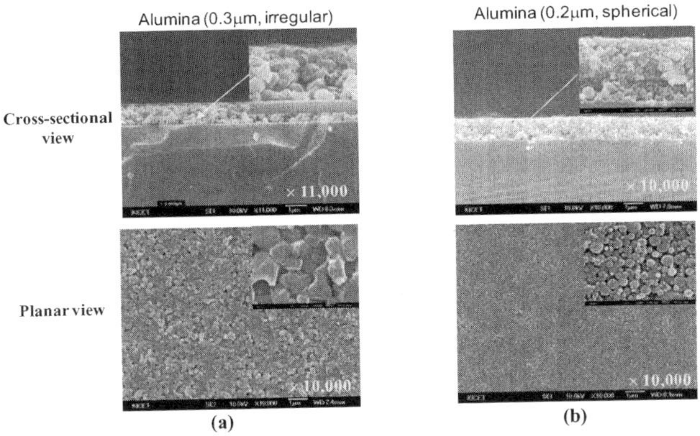

Figure 5 Cross sectional and planar view of the microstructures of inkjet printed alumina powder bed (as dried): (a) non-spherical alumina powders and (b) spherical with multi-modal size and distribution.

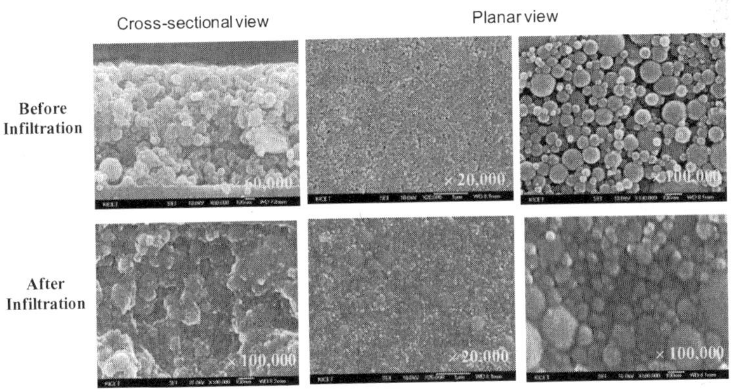

Figure 6 Cross sectional view of the resin (BCB) infiltrated alumina layer: (a) before infiltration and (b) after infiltration.

Table 1 Powder packing densities with particle size and shapes of alumina powders

Powders	No. of jetting layers	Thickness (um)	Area (mm^2)	Packing density
D_{50} = 0.3um Irregular shape	5	5.12	149.32	57.6%
D_{50} = 0.2um Spherical shape	4	5.23	138.53	68.5%

Table 2 Dielectric properties of alumina thick films with different polymer resins

Resins	Curing temp. (°C)	Curing time (hrs)	Q factor (@1MHz)
Epoxy	270	5	64 ± 8
BCB	150	5	130 ± 41
PPO	300	5	248 ± 34

Alumina-Polyimide Composite Thick Films by Aerosol Deposition Method

Applying the aerosol deposition method (ADM) is another approach to form a dense ceramic film at low temperature, actually at room temperature. We used alumina powders and polyimide powders in this experiment as raw materials for low loss dielectric film. Figure 9 illustrates the schematics of aerosol deposition process. A mixed gas consisted of He and O_2 was used as a carrier gas in the nozzle. Two types of alumina powders with different sodium impurity contents, normal sodium Al_2O_3 (A-161SG, Na_2O 0.18 wt%, Showa Denko) and low sodium Al_2O_3 (AL-160SG-3, Na_2O 0.06 wt%, Showa Denko) powders with particle sizes about 0.5um, were used in this experiment in order to compare the effect of impurity on the dielectric loss of alumina film.

The alumina powders were heat treated up to 900°C (pre-annealing) to examine the variation of microstructure and dielectric properties (Figure 10 and 11). The surface of an as-deposited alumina film has an un-even surface morphology as shown in Figure 10, and the rough surface was flattened by using the 900°C annealed powders. The dielectric loss of low sodium content alumina film was much lower than the films with normal sodium contents and the loss factor was further decreased with annealing temperature as shown in Figure 11. We found that lowering the dielectric losses were attributed to the microstructure enhancement due to crystallinity increase as well as sodium content decrease with annealing.

Further enhancement of dielectric properties of alumina films are obtained by mixing polyimide powders about 3 wt%. Polyimide (PI) powders were pre-milled for downsizing the particle sizes to 3um. The microstructures and dielectric losses of alumina and alumina/PI composite films are compared as shown in Figure 12 and 13, respectively. The crystalline size of alumina/PI composite film was larger (smaller than 50nm) than that of alumina film (about 200nm) which attributed to the polyimide addition. Figure 13 and Table 3 shows the dielectric properties of alumina, alumina/PI composite, and PI films. The dielectric constant and loss were decreased by low-K PI addition (K = 4.25 @1MHz) and larger alumina crystal size in the film. The addition of PI also improved the brittleness of alumina film.

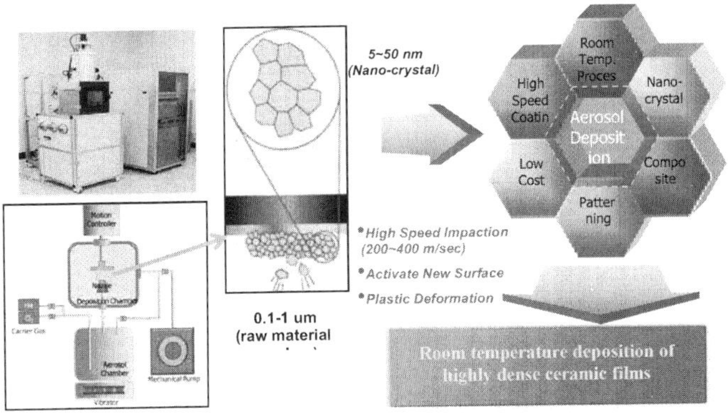

Figure 7 Schematics of aerosol deposition method (ADM) process.

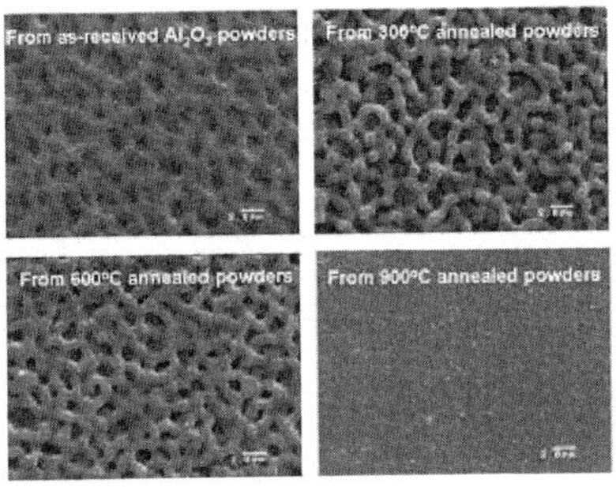

Figure 8 Surface morphologies of low sodium alumina film by ADM.

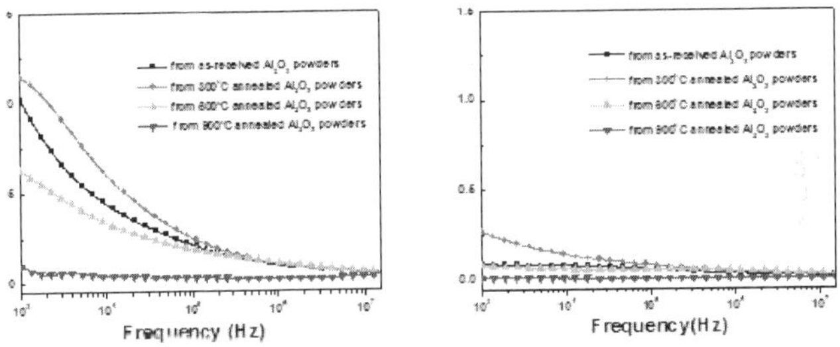

(a) *(b)*

Figure 9 Variation in the dielectric losses after heat treatment at different temperatures : (a) Normal sodium Al_2O_3 (A-161SG) and (b) Low sodium Al_2O_3 (AL-160SG-3) powders.

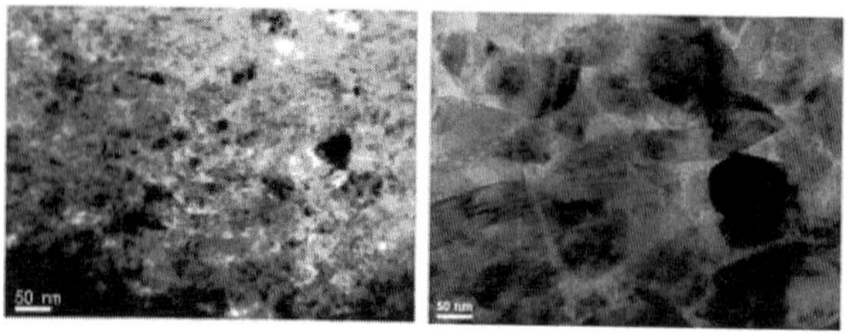

(a) (b)

Figure 10 TEM images of (a) Al_2O_3 and (b) Al_2O_3-PI composite films.

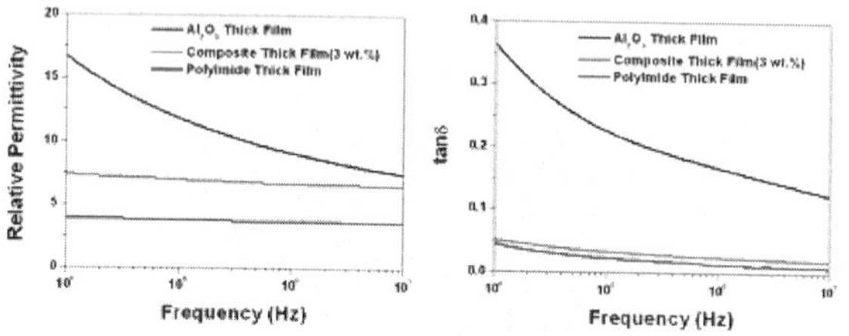

Figure 11 Variation in the dielectric properties of Al_2O_3-PI composite films by ADM.

Table 3 Summary of the dielectric properties of AD films with source materials

Film materials	Dielectric constant (@ 1 MHz)	Loss tangent (@ 1 MHz)
Al_2O_3	9.49	0.0095
Al_2O_3-PI composite	9.02	0.0071
Polyimide (PI)	4.25	0.0067

SUMMARY

Ceramic-organic composite thick films over 50 vol. % of dielectric powder loading has been fabricated at the temperatures lower than 300°C using several processing technologies such as ink-jet printing and aerosol deposition method. High solid loading of dielectrics with submicron size powders up to 68 vol. % was obtained by ink-jet printing via minimized polymer vehicles and controlled particle shape, size and distribution. Dielectric properties of thus obtained alumina-PPO resin composite film exhibited dielectric constant of 4.0 – 4.6 and Q factor of 248±33 at 1MHz. Alumina-polyimide composite film was fabricated by aerosol deposition method, and the dielectric constant and loss value were lower than the film made of alumina itself.

ACKNOWLEDGMENTS

This work was supported by a grant from the Fundamental R&D Program for Core Technology of Materials funded by the Ministry of Knowledge Economy (Grant number M2007010011).

REFERENCES

[1] J.H. Song, M.J. Edirisinghe, J.R.G. Evans, *J. Am. Ceram. Soc.*, 82 (12) (1999).

[2] J. Akedo and M. Lebedev, *Jpn. J. Appl. Phys.*, vol. 38, part 1, No.9B, 5397 (1999).

[3] N. Ramakrishnan, P.K. Rajesh, P. Ponnambalam, K. Prakasan, *J. Mat. Proc. Tech.*, 169 (2005) 372-38.

[4] K. McGeary, *J. Am. Ceram. Soc.*, 44 (10), 513-522 (1961).

[5] E. H. Koo, Y.H. Son, H.W. Jang, H.T. Kim, Y.J. Yoon, J.H. Kim, *Ceramic Transactions* (The American Ceramic Society, MS&T 2008, Pittsburg).

EFFECT OF RARE EARTH ELEMENTS DOPING ON THE ELECTRICAL PROPERTIES OF $(Ba,Sr)TiO_3$ THIN FILM CAPACITORS

N. Kamehara
Fujitsu Quality Laboratory LTD., Kawasaki, Kanagawa, Japan

K. Kurihara
Fujitsu Laboratories LTD., Atsugi, Kanagawa, Japan

ABSTRACT

Barium strontium titanate (BST) thin film capacitors are being intensively investigated for tunable microwave devices, because of their high permittivity, low dielectric loss in the microwave region and field dependent permittivity. This study investigates the effect of rare earth elements doping on the electrical properties of BST thin film capacitors.

BST thin films were deposited by an RF magnetron sputtering technique on Si wafers. BST films were prepared with Y concentration of 0-5%. Lattice parameters were measured using synchrotron radiation X-ray analysis.

The results show that Y doped BST capacitors exhibit not only significantly higher permittivity but also low leakage current density as compared to nominally undoped BST capacitors. X-ray diffraction results show the film strain state strongly depends on film composition and dopants with tunability decreasing with increasing tensile strain.

INTRODUCTION

Barium strontium titanate (BST) thin films have a great potential for the device applications in tunable microwave devices, where capacitors with large voltage tunability, low dielectric loss, and low leakage currents are required as well as semiconductor memory devices [1-3]. Recently, many groups have researched the effect of stress and strain on the dielectric constant (k) of BST thin films, which have lower k as compared to the bulk [4-9]. Many reports have investigated the modification of the elastic strain in epitaxial films by using different substrates [4-7] or by adjusting film thickness [4]. Doping of foreign elements is another efficient method for improving the electrical properties of BST-based thin film capacitors. As with compositional changes achieved through adjusting the Ba/Sr ratio [8], lattice distortion in thin BST films should also be modified by doping.

In this research, we studied the Y incorporation and its effects on the dielectric behavior of BST thin films.

EXPERIMENTAL

The BST thin films in this study were deposited using an off-axis sputtering system. First, a blanket bottom Pt electrode was deposited on SiO_2/Si wafer with a TiO_2 adhesion layer. Subsequently, BST thin films were deposited using a RF sputtering method at 520 ^{o}C. To investigate the electrical properties of BST thin films with different Y doping concentrations, BST ceramics targets with different Y doping concentration were used. The capacitors used for electrical characterization were fabricated using plasma etching after top Pt electrode deposition. BST thin films and bottom electrodes were deposited in series without breaking the vacuum. Low frequency (100 Hz) capacitance characteristics were measured by an HP4194A impedance gain phase analyzer with ac oscillation level of 50 mV. Leakage currents were recorded using a

HP 4339B high resistance meter and HP4156C precision semiconductor parameter analyzer. The crystallinity of the films was investigated using X-ray diffraction in θ - 2 θ and ϕ scan.

RESULTS AND DISCUSSION

Figure 1 shows an example of cross-sectional TEM image of the sample structure. The films with different Y concentrations exhibit columnar microstructures with clean grain boundaries; no voids or second phases were observed.

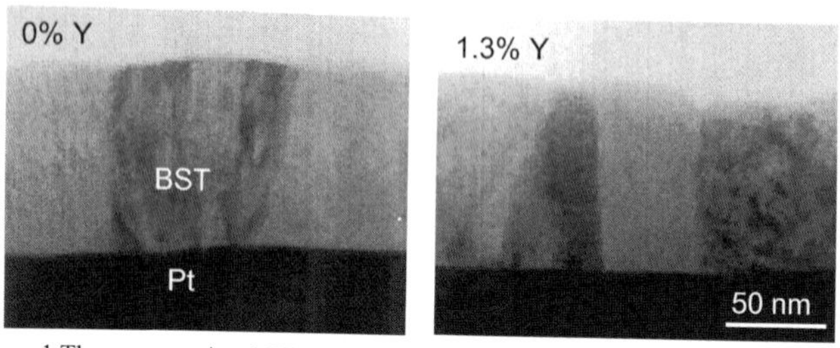

Figure 1 The cross-sectional TEM images of the sample structures without top electrodes

The zero-field permittivity as a function of Y dopant concentration was plotted in Fig. 2. The permittivity increases as Y content increases and reaches the maximum at Y percentage equal to 1.3%.

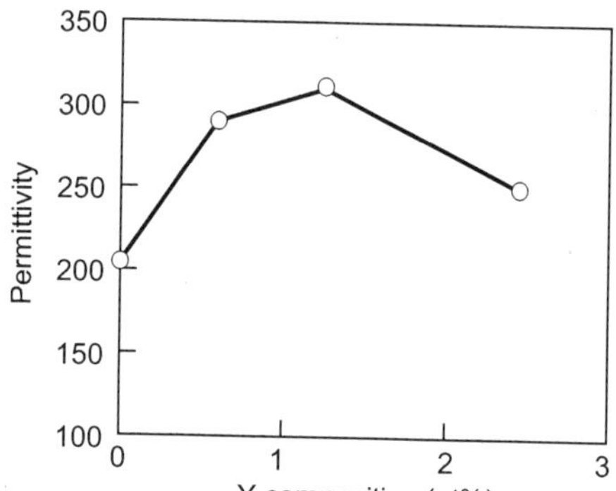

Figure 2 Zero-field permittivity as a function of Y composition

Figure 3 shows the permittivity versus electrical field for the films with Y composition ranging from 0% to 1.3%. In particular, the maximum permittivity of the 1.3%-Y doped BST thin films is 70% higher than that of the undoped BST thin films, while maintaining the dielectric loss lower than 1%. Furthermore, 1.3% Y-doped BST thin films also show much greater tunability than that of undoped BST thin films.

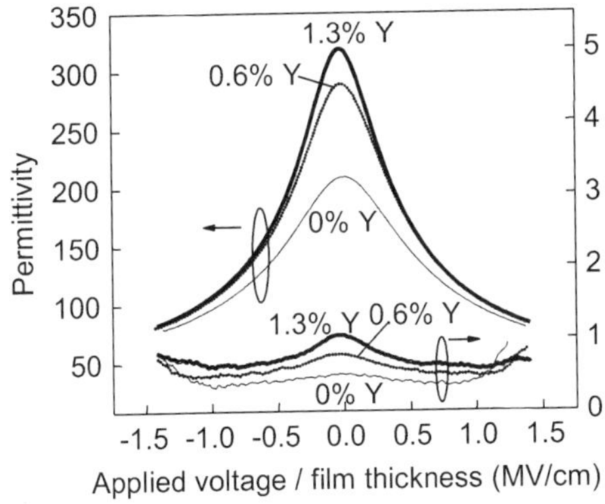

Figure 3 Permittivity and loss tangent as a function of electrical fields

The J-V data shown in Fig. 4 shows that by Y-doping, the leakage current density at room temperature decreases by as much as a factor of 10 at both positive and negative polarity. It is well known that Y dopants compensate charged defects in bulk Perovskite oxide ceramics. The reason of this excellent Y doping effect on leakage property may be charge compensate as same as bulk ceramics.

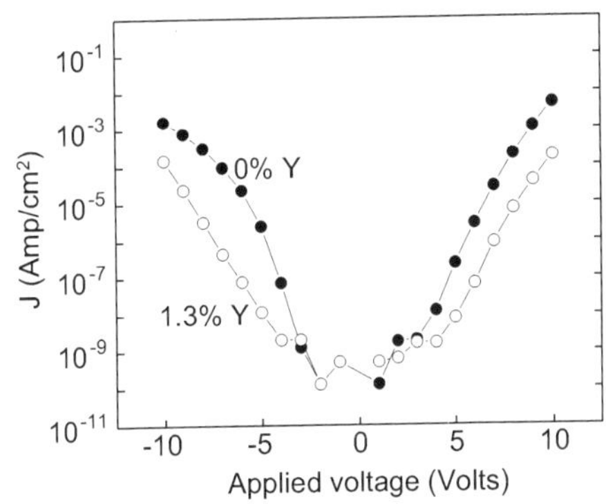

Figure 4 J-V characteristics of undoped and Y doped BST thin film capacitors

To further investigate the strain effects, $\sin^2\chi$ analysis was applied to evaluate the elastic strain in the films with different Y-dopant concentrations. As a first attempt, the unstressed lattice parameter was estimated by assuming the film with (100) anisotropic orientation texture. Fig. 5 shows the the elastic strain for BST films with different composition. As the Y concentration increases, the elastic strain decreases and reaches the minimum when Y concentration is equal to 1.3%. It seems that Y doping is effective on relaxing residual tensile strain in BST thin films.

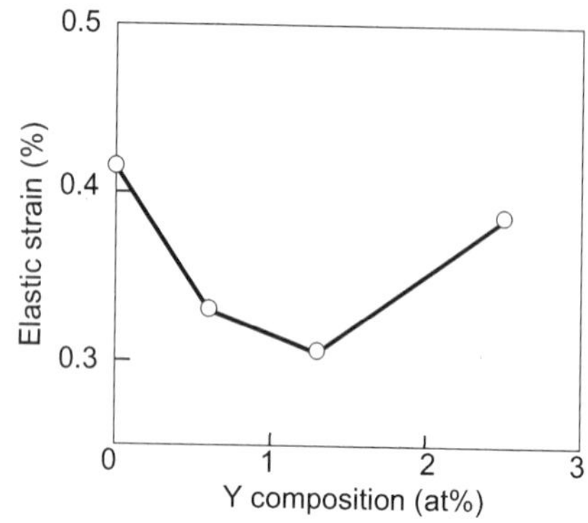

Figure 5 Elastic strains as a function of Y composition

The zero-field permittivity is re-plotted as a function of elastic strain from the films with different Y composition in Fig. 6, showing a clear trend that the permittivity decreases with increasing tensile strain. There are many reports about the influence of strain on permittivity [9, 10]. Those results indicate that Y doping relaxes residual tensile strain in the BST thin films and as the results, permittivity increases with Y composition.

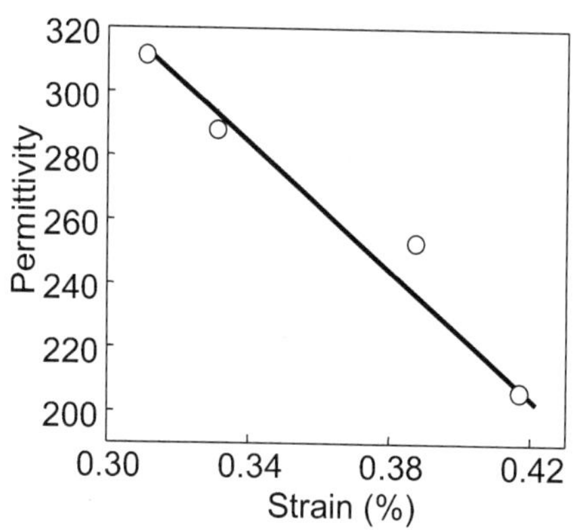

Figure 6 Zero-field permittivity as a function of elastic strain

CONCLUSION

Electrical properties of Y doped sputter deposited BST thin films were investigated for the tunable microwave device applications. Y-dopant greatly increases the permittivity and tunability of BST thin films. Furthermore, Y-doping of 1.3% decreases the leakage current by more than one order of magnitude. High performance BST thin films can be achieved through simultaneous optimization of strain and defect chemistry by Y doping.

REFERENCES

[1] J. D. Baniecki, T. Shioga, and K. Kurihara, Microstructural and electrical properties of $(Ba_x Sr_{1-x})Ti_{1+y} O_{3+z}$ thin films prepared by rf magnetron sputtering, *Integr. Ferroelectr.*, **46**, 221 - 232 (2002).
[2] C. S. Hwang, S. O. Park, H. J. Cho, C. S. Kang, H. K. Kang, S. I. Lee, and M. Y. Lee, Deposition of extremely thin $(Ba,Sr)TiO_3$ thin films for ultra-large-scale integrated dynamic random access memory application, *Appl. Phys. Lett.*, **67**, 2819 - 2821 (1995).
[3] D. Dimos and C. H. Mueller, Perovskite thin films for high-frequency capacitor applications, *Annu. Rev. Mater. Sci.*, **28**, 397 - 419 (1998).
[4] Z. G. Ban and S. P. Alpay, Phase diagrams and dielectric response of epitaxial barium strontium titanate films: A theoretical analysis, *J. Appl. Phys.*, **91**, 9288 - 9296, (2002).
[5] N. A. Pertsev, A. G. Zembilgotov, and A. K. Tagantsev, Effect of mechanical boundary conditions on phase diagrams of epitaxial ferroelectric thin films, *Phys. Rev. Lett.*, **80**, 1988 - 1991 (1998).
[6] A. Sharma, Z. G. Ban, S. P. Alpay, and J. V. Mantese, The role of thermally-induced internal stresses on the tunability of textured barium strontium titanate films, *Appl. Phys. Lett.*, **85**, 985 - 987 (2004).
[7] T. R. Taylor, P. J. Hansen, B. Acikel, N. Pervez, R. A. York, S. K. Streiffer, and J. S. Speck, Impact of thermal strain on the dielectric constant of sputtered barium strontium titanate thin films, *Appl. Phys. Lett.*, **80**, 1978 - 1980 (2002).
[8] B. H. Park, E. J. Peterson, Q. X. Jia, J. Lee, X. Zeng, W. Si, and X. X. Xi, Effects of very thin strain layers on dielectric properties of epitaxial $Ba_{0.6}Sr_{0.4}TiO_3$ films, *Appl. Phys. Lett.*, **78**, 533 - 535 (2001).
[9] T. M. Shaw, Z. Suo, M. Huang, E. Liniger, R. B. Laibowitz, and J. D. Baniecki, The effect of stress on the dielectric properties of barium strontium titanate thin films, *Appl. Phhys. Lett.*, **75**, 2129 - 2131 (1999).
[10] W. Chang, A. M. Gilmore, W.-J. Kim, J. M. Pond, S. W. Kirchoefer, S. B. Qadri, D. B. Chirsey, and J. S. Horwitz, Influence of strain on microwave dielectric properties of $(Ba,Sr)TiO_3$ thin films, *J. Appl. Phys.*, **87**, 3044 - 3049 (2000).

MICROWAVE PROCESSING OF DIELECTRICS FOR HIGH POWER MICROWAVE APPLICATIONS

Isabel K. Lloyd[1,2], Yuval Carmel[2], Otto C. Wilson, Jr.[3], and Gengfu Xu[4]

[1]Materials Science and Engineering and [2]Institute for Research in Electronics and Applied Physics
University of Maryland, College Park, MD, USA

[3]Department of Biomedical Engineering
Catholic University
Washington D.C., USA

[4]Fuel Cell Energy, Inc
Danbury CN USA

ABSTRACT

While most electronic applications are moving rapidly towards solid-state devices, high power microwave communication devices still utilize vacuum tube technology to achieve the required power density. Traditionally BeO and BeO composites were used as support and tuning dielectrics. However, AlN based materials are now desirable due to the health effects of BeO. Dielectric constant and thermal conductivity are the key properties for high power communications. Even with liquid phase sintering it is not easy to achieve high thermal conductivity in AlN materials using conventional firing. Microwave firing promotes rapid development of thermal conductivity. With microwave firing, thermal conductivities above 200 W/mK in pure AlN can be achieved in four hours. In addition, thermal conductivities of 100-150 W/mK can be achieved in tailored dielectric constant AlN-TiB2 and AlN-SiC tuning dielectric composites in 1-2 hours. Our approach is applicable to other applications for AlN and its composites.

INTRODUCTION

While AlN is electrically insulating, it is a good thermal conductor. This combination is desirable for high power electronic packaging and support and tuning dielectrics (attenuators and terminators) for high power density microwave communication tubes. In particular, AlN and its composites are of interest as replacement materials for BeO and BeO-SiC composites in microwave communication tubes since they do not have the adverse health effects associated with beryllia and berylliosis. The goal of our program was to explore microwave processing of AlN and AlN-TiB_2 and AlN-SiC composites and determine if they were good alternatives to BeO and BeO-SiC microwave tube materials. In particular, we were interested in the ability of microwave sintering to densify high thermal conductivity, tailored dielectric loss materials from standard, commercially available powers without extended annealing.

The program took advantage of the computerized microwave processing system we developed including simulation of thermal and dielectric behavior as a function of microstructure development during sintering to prevent thermal runaway [1,2]. It also led to the development of a unique, reusable high temperature BN, ZrO_2 insulation system [3].

EXPERIMENTAL METHODS

A computerized, highly overmoded 2.45 GHz, 3 kW microwave furnace with an advanced dynamic proportional–integral–differential (PID) feedback loop and optical temperature measurement was used for all the processing experiments. As noted above, simulation of the densification behavior was incorporated into the control system to prevent thermal overrun or runaway. This was especially important for the lossy composites with SiC and TiB_2. The data for the simulations and calibration of the optical pyrometers was determined from preliminary runs with dummy samples. The furnace was evacuated and then back filled with atmospheric pressure Ar or N_2 gas for the experiments.

Table 1 lists the materials and compositions used in the study. Samples were 2.5 cm disks uniaxially pressed at 35 MPa and then isopressed at 300 MPa. Samples were embedded in AlN powder in a BN crucible before being placed in the insulation casket and fired at temperatures from 1760°C to 2100°C for 30 minutes to 4 hours depending on the sample composition. Figure 1 indicates representative firing conditions.

Fired samples were characterized using immersion density, thermal conductivity (laser flash technique), x-ray diffraction, microwave permittivity and loss tangent, scanning electron microscopy of fractured and polished surfaces and Vickers hardness measurements to determine hardness and relative indentation toughness.

Table 1: Materials

Material	Source	Characteristics	Purpose
AlN	Advanced Refractory Technology, Buffalo NY	Grade 100, mean particle size 4.5 μm	High thermal conductivity matrix
TiB_2	HC Stark	Grade F, mean particle size 2.0-3.5 μm	Lossy second phase (0, 15, 30, 38 wt%)
SiC	ESK Aldrich	Mean particle size 1-3μm Mean particle size 6-10 μm	Lossy second phase (0, 20, 40 wt%)
Y_2O_3	Molycorp	99.99%	Liquid phase sintering additive (composite samples-4 wt%; AlN-6 wt%)

RESULTS AND DISCUSSION

While single crystal AlN has a theoretical thermal conductivity of 319 W/mK [4], polycrystalline AlN typically has much lower thermal conductivity due to grain boundaries and second phases at the grain boundaries resulting from the oxidation of the AlN to form aluminum oxide and the reaction the aluminum oxide with liquid phase sintering additives like yttrium oxide. Typical values for pressureless sintered materials range from 70-180 W/mK when the study was begun [5]. Higher values of thermal conductivity required long anneals to eliminate oxygen and second phases. For example Watari et al [6] were able to produce samples with a thermal conductivity of 272 W/mK by annealing for 100 hours at 1900°C.

Figure 1 shows the room temperature thermal conductivity of our AlN and AlN composites as a function of composition and firing temperature, atmosphere and time. Note that with microwave sintering, thermal conductivity in excess of 200W/mK is possible with firing for only 2 hours to 99.5% theoretical density. When the firing time was increased to 4 hours, thermal conductivity increased to 225 W/mK. X-ray diffraction showed that the lattice parameter increased as firing time was increased to 2 hours, indicated a decrease in oxygen content. However, when firing times were increased beyond 2 hours, the lattice constant changed very little but the amount of second phase decreased substantially. This indicates that microwave sintering is beneficial in developing high thermal conductivity AlN.

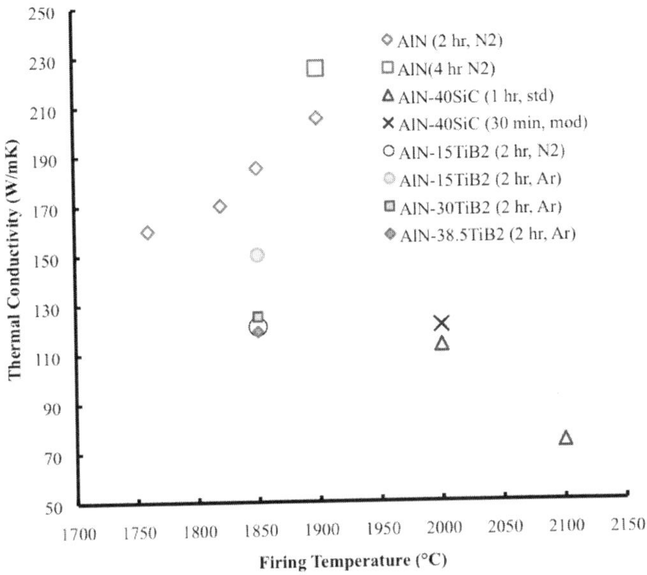

Figure 1: Thermal conductivity as a function of firing temperature, composition, firing time and atmosphere.

Figure 1 also shows that microwave processing is helpful in producing AlN matrix composites with high thermal conductivities. In particular, we found that we were able to produce composites with thermal conductivities close to what would be predicted using dielectric rules of mixtures such as the Eshelby inclusion model [7]. As with AlN, the thermal conductivity of the composites was strongly related to the microstructure. It increased with increasing AlN grain to grain connectivity and decreased with the amount of second phases, oxygen and solid solution formation. Second phase formation was a particular problem with the AlN-TiB_2 composites. Firing in nitrogen lead to the formation of BN as a second phase. It also enhanced the formation of TiN. Firing in argon inhibited the formation of BN and limited the formation of TiN. Solid solution formation was a problem for the AlN-SiC composites. When they were fired using conventional heating rates, solid solution formation was evident in x-ray diffraction and hardness measurements as well as difficulty locating the SiC grains in the microstructures of composites made with the smaller ESK SiC.

Figure 2 shows the Vickers Hardness as function of SiC composition and where the hardness was measured for samples with the larger Aldrich powder fired with standard heating rates. The hardness near the rim of the grains is much lower due to solid solution formation with the much softer AlN. The Vickers Hardness measurements were done as a screening test for mechanical properties since AlN composites are also of interest as higher temperature structural composites. However, they provide an instructive link between the microstructure and the thermal and mechanical properties.

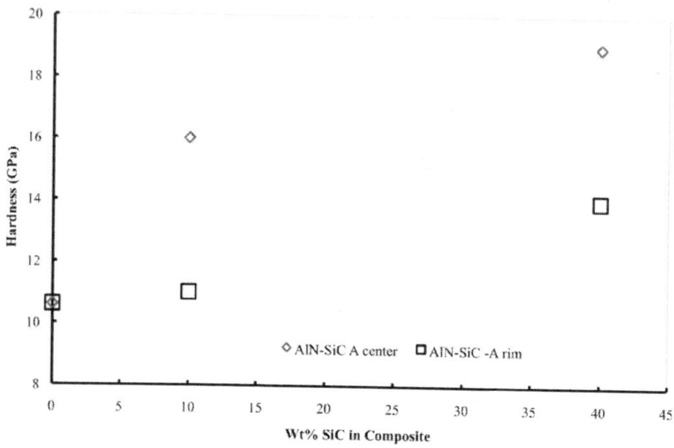

Figure 2: Vickers Hardness as a function of microstructural location for composites with Alrich SiC fired at 2000°C for 1 hour.

Figure 3 shows the dielectric response of our composites at microwave frequencies. It shows that the dielectric properties can be tailored using composition. Note in particular that while the AlN-15TiB_2 and AlN-20SiC composites have very similar relative dielectric constants, the SiC composite exhibits a higher loss. The ability to tailor loss and dielectric constant was critical to showing that AlN composites could serve as high thermal conductivity replacements for BeO-SiC composites. The dielectric measurements are for samples are for samples that were not optimized for thermal conductivity. It is expected that the dielectric properties would change at least slightly if the samples were fired longer to increase their thermal conductivity.

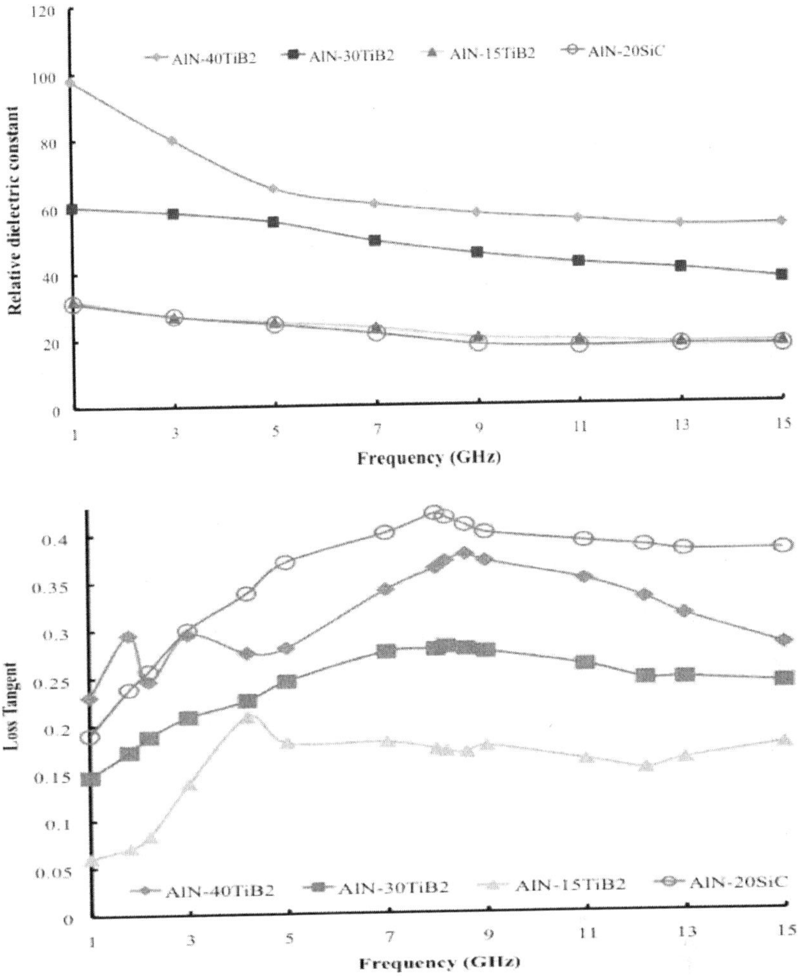

Figure 3: Relative dielectric and loss tangent for AlN composites. Note, AlN-TiB_2 samples were fired for 2 hours at 1850°C in Ar, the AlN-SiC sample was fired for 30 minutes at 2000°C.

CONCLUSIONS

Microwave liquid phase sintering useful approach to develop high thermal conductivity microwave communication tube dielectrics with tailored dielectric constant and loss tangent. Microwave sintering appears to enhance development of thermal properties and allow densification without extensive solid solution formation in composites. While, this was a largely a feasibility study so that processing parameters were not explored extensively, AlN samples with 225 W/mK were produced with a four hour densification cycle at 1900°C.

ACKNOWLEDGEMENTS

Financial support was provided by the U.S. Naval Research Laboratory, Vacuum Electronics Division and the AFOSR MURI 99 Program on Microwave Vacuum Electronics. We thank Tayo Olorunyolemi and Amikam Birnboim for their intellectual and experimental contributions to the program.

REFERENCES

1. A Birnboim , T Olorunyolemi and Y Carmel. "Calculating the thermal conductivity of heated powder compacts," J Am Ceram Soc 84:6 1315-1320 (2001).

2. A Birnboim , and Y Carmel. "Simulation of microwave sintering of ceramic bodies with complex geometry," J. Am Ceram Soc 82:11 3024-3030 (1999).

3. GF Xu, T Olorunyolemi, Y Carmel . IK Lloyd and OC Wilson, "Design and construction of insulation configuration for ultra-high-temperature microwave processing of ceramics," J Am. Ceram. Soc 86:12 2082-2086 (2003).

4. A AlShaikhi and G P Srivastava, "Role of additives in enhancing the thermal conductivity of AlN ceramics," J Phys D-Appl Phys 41:18 Article Number: 185407 (SEP 21 2008).

5. GF Xu, T Olorunyolemi, OC Wilson, Jr., IK Lloyd, Y Carmel, "Microwave sintering of high-density, high thermal conductivity AlN," J MATERIALS RESEARCH 17 (11): 2837-2845 2002

6. K Watari, H Nakano, K Urabe, K Ishizaki, SX Cao SX and K Mori, "Thermal conductivity of AlN ceramic with a very low amount of grain boundary phase at 4 to 1000 K," J Mater Res 17:11 2940-2922 (2002).

7. JP Calame and D Abe, "Applications of Advanced Materials Technologies to Vacuum Electronic Devices," Proc. IEEE 87:5 840-864 (1999).

FERROELECTRIC DOMAINS IN LEAD FREE PIEZOELECTRIC CERAMICS

Toshio Ogawa
Department of Electrical and Electronic Engineering, Shizuoka Institute of Science and Technology
Fukuroi, Shizuoka 437-8555, Japan

Masahito Furukawa
Materials and Process Development Center, TDK Corporation
Narita, Chiba 286-8588, Japan

ABSTRACT

DC poling field dependence of piezoelectricity was investigated to evaluate the domain structures in lead free ceramics composed of (1-x)(Na,K,Li,Ba)($Nb_{0.9}Ta_{0.1}$)O_3-x$SrZrO_3$ (x=0-0.07) comparing with the structures of PZT, $PbTiO_3$ (PT) and $BaTiO_3$ (BT) ceramics. Poling was conducted at 150℃ for 30 min while varying the poling field (E) between ±4.0 kV/mm. Increasing x from 0 to 0.05, the relative dielectric constant (ε_r), electromechanical coupling factor of planar mode (k_p), frequency constant of k_p mode (fc_p) and piezoelectric strain constant (d_{33}) vs E show typical domain clamping at a specific E. The change in the ε_r, k_p, fc_p and d_{33} vs E became small at x=0.06-0.07, because of the approaching to the paraelectric phase. The ceramics at x=0 show typical ε_r, k_p, fc_p and d_{33} vs E such as PT and BT ceramics. The maximum k_p (48%) and d_{33} (307 pC/N) were obtained at x=0.05 with lowest fc_p of 2964 Hz·m while appearing ε_r, k_p, fc_p and d_{33} vs E such as typical tetragonal hard PZT ceramics. Since the domain orientation in the ceramics was accompanied with deformation of the crystals while applying E, it was clarified that large k_p and d_{33} in the lead free ceramics needed to realize low fc_p which corresponds to low Young's modulus.

INTRODUCTION

Material research regarding lead free piezoelectric ceramics has been paid much attention because of global environmental considerations. The key practical issue is the difficulty to realized large piezoelectricity such as electromechanical coupling factors and piezoelectric strain constants. A planar coupling factor of disk (k_p) is closely related to the orientation degree of ferroelectric domains through the DC poling process. We had already reported the behaviour of domain orientation in PZT[1-5], $PbTiO_3$ (PT)[6] and $BaTiO_3$ (BT)[7] ceramics and a relaxor single crystal of Pb[($Zn_{1/3}Nb_{2/3}$)$_{0.91}$$Ti_{0.09}$]$O_3$ (PZNT)[8] by measuring the piezoelectricity vs DC poling field. In this study, the poling characteristics, especially DC poling field dependence of dielectric and piezoelectric properties were investigated in lead free ceramics comparing with the characteristics of PZT, PT and BT ceramics.

EXPERIMENTAL PROCEDURE

The lead free ceramics evaluated are composed of (1-x)(Na,K,Li,Ba)($Nb_{0.9}Ta_{0.1}$)O_3-x$SrZrO_3$ (x=0-0.07) with small amount of MnO. The ceramics were fabricated by a conventional ceramic manufacturing process such as firing conditions of 1100～1200℃ for 2 hr. The DC poling temperature and the time are fixed at 150℃ and 30 min applied to the ceramic disk (dimensions: 14 mm$^{\phi}$ x 0.5 mm^{T}) with Ag electrode while varying the DC electric fields gradually (0.2/0.25/0.5 kV/mm each) from E=0→+5.0→0→ -5.0→0 to +5.0 kV/mm. After each poling, the dielectric and piezoelectric properties were measured at room temperature using an LCR meter (HP4263A) and an impedance/gain-phase analyzer (HP4194A). P-E hysteresis loops were observed by applying a bipolar triangle pulse, the pulse period of which was 400 ms.

RESULTS AND DISCUSSION

Dielectric and Piezoelectric Properties

Figures 1(a)~(d) illustrate the relationships among the $SrZO_3$ (SZ) composition (x) vs relative dielectric constant (ε_r), planar coupling factor (k_p), frequency constant (fc_p), and piezoelectric strain constant (d_{33}) in (1-x)(Na,K,Li,Ba)($Nb_{0.9}Ta_{0.1}$)O_3-x$SrZrO_3$ (x=0-0.07) with small amount of MnO. In this case, the DC poling was conducted at 150℃ for 30 min by applying E of 4.0 kV/mm.

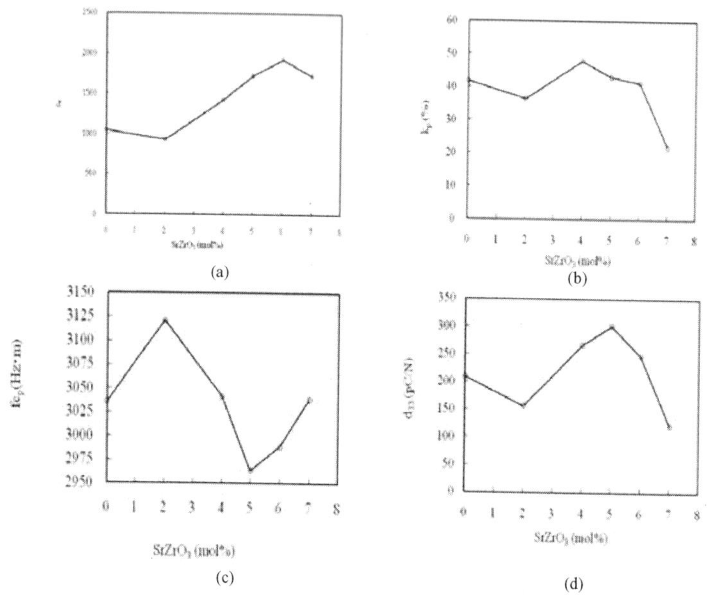

Figure 1. $SrZrO_3$ composition (mol%) dependence of (a) relative dielectric constant (ε_r), (b) planar coupling factor (k_p), (c) frequency constant (fc_p), and (d) piezoelectric strain constant (d_{33}) in (1-x) (Na,K,Li,Ba)($Nb_{0.9}Ta_{0.1}$)O_3-x$SrZrO_3$ (x=0-0.07) with small amount of MnO.

Although the ε_r slightly decreased at x=0.02, the ε_r increased with x, and the maximum ε_r of 1931 was obtained at x=0.06. From the observation of P-E hysteresis loops, since the loop at x=0.07 shows the ferroelectricity, it became slim in compassion with the ones at x=0.04 [Figs. 2(a) and 2(b)]. Therefore, the compositions over x=0.06, the ceramic crystal phase approaches to paraelectric phase with decreasing the ε_r and the Curie temperature. The maximum k_p of 48% was realized at x=0.04. The relationship between fc_p (half of bulk wave velocity) and x shows a reverse tendency in the case of k_p vs x. The maximum k_p composition region ($0.04 \leqq x \leqq 0.06$)

corresponds to the minimum fc_p composition region ($0.04 \leqq x \leqq 0.06$). It was considered that higher crystal orientation (domain alignment) by DC poling could be achieved under softening the ceramics, which mean to have low Young's modulus (low fc_p). The maximum d_{33} of 307 pC/N was obtained at x=0.05 because of the difference compositions to realize the maximum ε_r (x=0.06) and k_p (x=0.04).

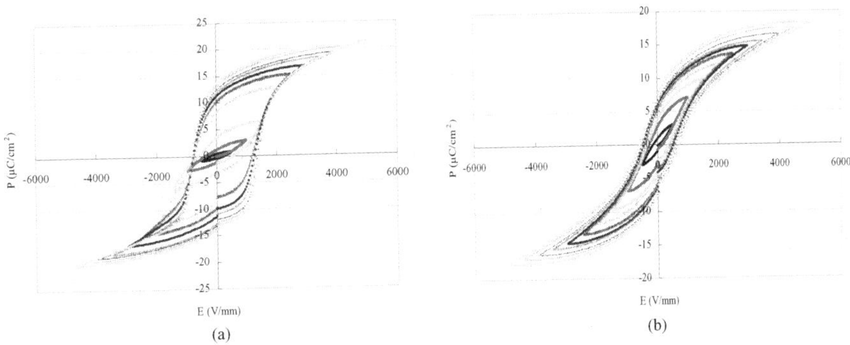

Figure 2. P-E hysteresis loops of (a) x=0.04 and (b) x=0.07 in (1-x) $(Na,K,Li,Ba)(Nb_{0.9}Ta_{0.1})O_3$-$xSrZrO_3$ with small amount of MnO at various applying fields. Maximum E is ±5.0 kV/mm.

Poling Field Dependence

Figures 3~6 show the effect of DC poling field (E) on ε_r, k_p and, fc_p and d_{33} at various SZ compositions of x when E was varied from 0 to ±4.0 kV/mm. Increasing x from 0 to 0.05, the relationships between the ε_r, k_p, fc_p and d_{33} vs. E show typical domain clamping at a specific E (Fig. 7).

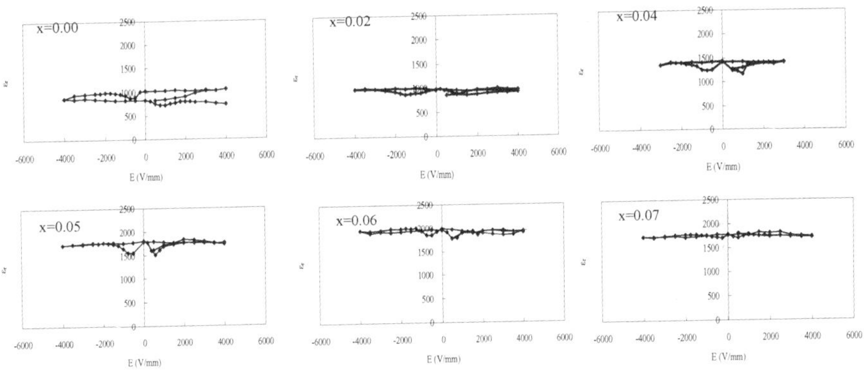

Figure 3. DC poling field dependence of ε_r at different compositions of x=0.00-0.07 in (1-x) $(Na,K,Li,Ba)(Nb_{0.9}Ta_{0.1})O_3$-$xSrZrO_3$ (x=0-0.07) with small amount of MnO.

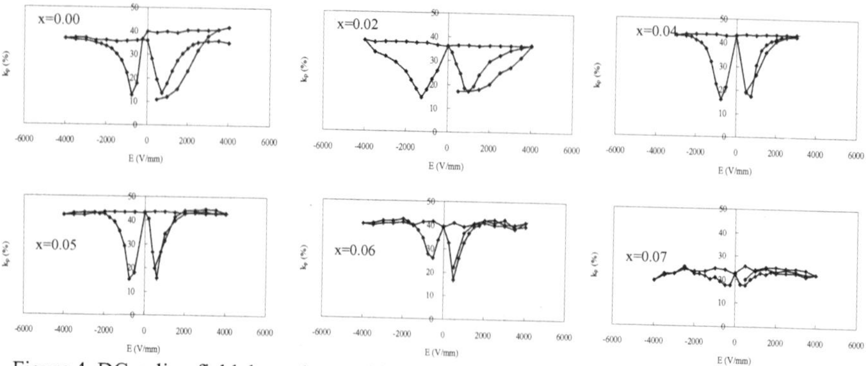

Figure 4. DC poling field dependence of k_p at different compositions of x=0.00-0.07.

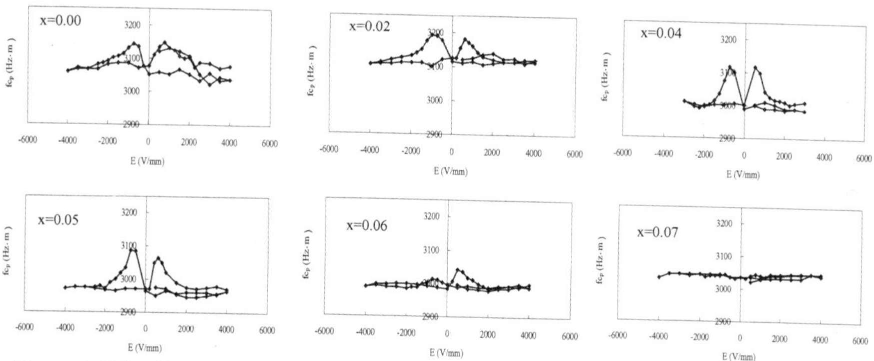

Figure 5. DC poling field dependence of fc_p at different compositions of x=0.00-0.07.

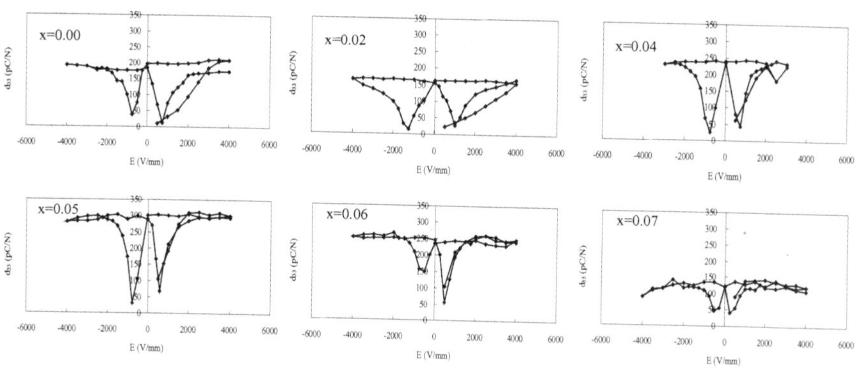

Figure 6. DC poling field dependence of d_{33} at different compositions of x=0.00-0.07.

It was considered that the minimum ε_r, k_p ,d_{33} and maximum fc_p owing to electrical domain clamping, such as ↑↓ [the arrow (↑) means domain orientation], occurred at the coercive fields corresponded to the specific E as mention previously.The change in the ε_r, k_p, fc_p and d_{33} vs E became small at x=0.06-0.07, because of the approaching to the paraelectric phase with abrupt decreasing k_p and d_{33}. The ceramics at x=0 show typical ε_r and fc_p vs E behaviour such as BT ceramics[7] and k_p vs E such as PT ceramics[6]. However, the ceramics at x=0.05 processed typical ε_r, k_p and fc_p vs E such as tetragonal PZT hard ceramics[1,3-5]. Moreover, the maximum k_p (48%) and d_{33} (307 pC/N) were obtained at the lowest fc_p of 2964 Hz·m at x=0.05. Since the domain orientation in the ceramics was accompanied with deformation of the crystal while applying DC poling field, it was clarified that large k_p and d_{33} in the lead free ceramics needed to realize low fc_p (low Young's modulus) in the ceramic compositions.

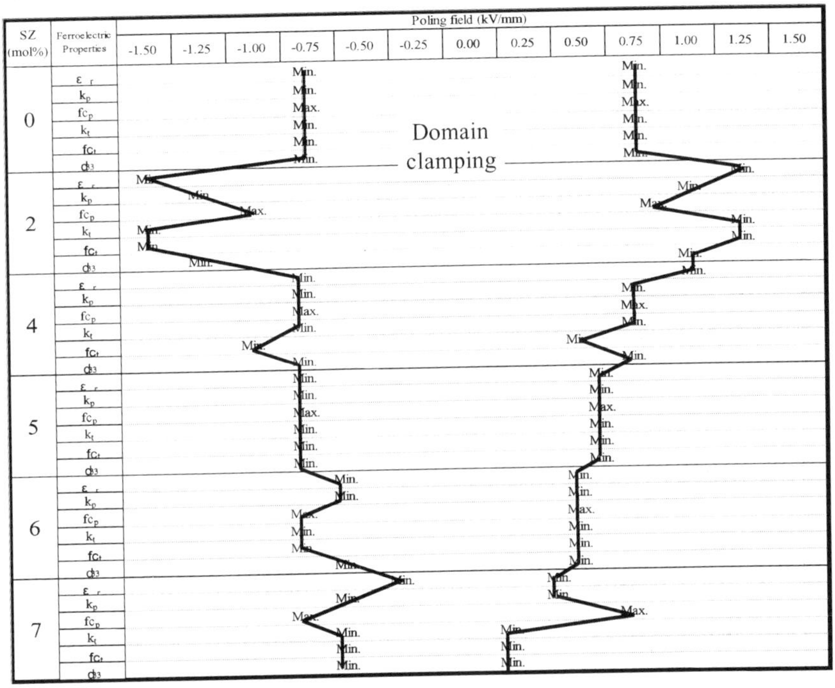

Figure 7. Relationships between DC poling field (E) to obtain minimum and maximum ε_r, k_p, fc_p, k_t*, fc_t*, d_{33} and $SrZrO_3$ (SZ) compositions of x=0.00-0.07. Domain clamping occurs at the same E to realize minimum ε_r, k_p, k_t, d_{33} and maximum fc_p. Typical domain clamping appeared in the compositions of x=0.00 and x=0.05.

* k_t is electromechanical coupling factor of thickness mode of disk, and fc_t is frequency constant of k_t mode.

CONCLUSIONS

DC poling field dependence of dielectric and piezoelectric properties were investigated in lead free ceramics in comparison with those in PZT, PT and BT ceramics. The domain clamping was observed in the lead free ceramics as well as in PZT, PT and BT ceramics. Higher electromechanical coupling factor could be obtained in the ceramics with the composition of low frequency constant which corresponds to low Young's modulus. Furthermore, the ceramics show the typical domain clumping because of easy deformation by DC poling field.

ACKNOWLEDGEMENTS

This work was partially supported by a Grant-in-Aid for Scientific Research (C) (No. 17560294) from the Ministry of Education, Culture, Sports, Science and Technology and the Research Foundation 2008 between the Academy and Industry of Fukuroi City.

REFERENCES

[1]T. Ogawa, A. Yamada, Y.K. Chung, and D.I. Chun, Effect of Domain Structures on Electrical Properties in Tetragonal PZT Ceramics, *J. Korean Phys. Soc.*, **32**, S724-S726 (1998).

[2]T. Ogawa and K. Nakamura, Poling Field Dependence of Ferroelectric Properties and Crystal Orientation in Rhombohedral Lead Zirconate Titanate Ceramics, *Jpn. J. Appl. Phys.*, **37**, 5241-45 (1998).

[3]T. Ogawa and K. Nakamura, Effect of Domain Switching and Rotation on Dielectric and Piezoelectric Properties in Lead Zirconate Titanate Ceramics, *Jpn. J. Appl. Phys.*, **38**, 5465-69 (1999).

[4]T. Ogawa, Domain Switching and Rotation in Lead Zirconate Titanate Ceramics by Poling Fields, *Ferroelectrics*, **240**, 75-82 (2000).

[5]T. Ogawa, Domain Structure of Ferroelectric Ceramics, *Ceramics International*, **26**, 383-390 (2000).

[6]T. Ogawa, Poling Field Dependence of Crystal Orientation and Ferroelectric Properties in Lead Titanate Ceramics, *Jpn. J. Appl. Phys.*, **39**, 5538-41 (2000).

[7]T. Ogawa, Poling Field Dependence of Ferroelectric Properties in Barium Titanate Ceramics, *Jpn. J. Appl. Phys.*, **40**, 5630-33 (2001).

[8]T. Ogawa, Poling Field Dependence of Piezoelectric Properties in Piezoelectric Ceramics and Single Crystal, *Ferroelectrics*, **273**, 371-376 (2002).

FABRICATION OF $SrTi_4Bi_4O_{15}$ PIEZOELECTRIC CERAMICS WITH ORIENTED STRUCTURE USING MAGNETIC FIELD-ASSISTED SHAPING AND SUBSEQUENT SINTERING PROCESSING (MFSS)

Satoshi Tanaka, Kazunori Mishina and Keizo Uematsu
Nagaoka University of Technology
1603-1 Kamitomioka, Nagaoka Niigata, 9402188 JAPAN

ABSTRACT

The magnetic field-assisted shaping and subsequent sintering processing (MFSS) is very promising for obtaining crystal oriented ceramics. This technique consists from two steps. First step is shaping process for green body with crystal oriented particles in a magnetic field. Second step is grain growth during sintering without the magnetic field. The target of this study was to fabricate strontium bismuth titanate ceramics with highly crystal-oriented structure and high density. The influence of particle size on the orientation structure was important for each fabrication step. The Lotgering factor of green sample, as an index of orientation degree, was slightly increased from 0.08 to 0.2 with particle size after shaping in strong magnetic field 10 Tesla. After sintering, the oriented structure drastically developed. The Lotgering factor increased up to about 0.6 from 0.2 of green sample, in which the particles with mean diameter 0.8-1.0μm were used as the starting materials. On the contrary, the relative density decreased with particle size from 98 to 96%. When using large particles with 0.8μm, both of the orientation degree and relative density showed high level.

INTRODUCTION

The magnetic field-assisted shaping and subsequent sintering processing (MFSS) method is reasonable for designing crystal oriented ceramics[1-17]. We have applied this method to alumina[8], titania[9], bismuth titanate family[10,11], and tungsten bronze system[16,17] etc. The advantage of the technique is the possibility of using conventional fine particles with near-spherical shape. They allow easy densification in subsequent sintering processes. The other advantage is the possibility of preferred orientation by changing direction of magnetic field according to their property.

Alternative materials for $Pb(Zr,Ti)O_3$ ferroelectrics are expected by increasing concern for the environmental problem[19-21]. Candidates for lead-free piezoelectric materials need the crystal oriented structure for improving their property, because they

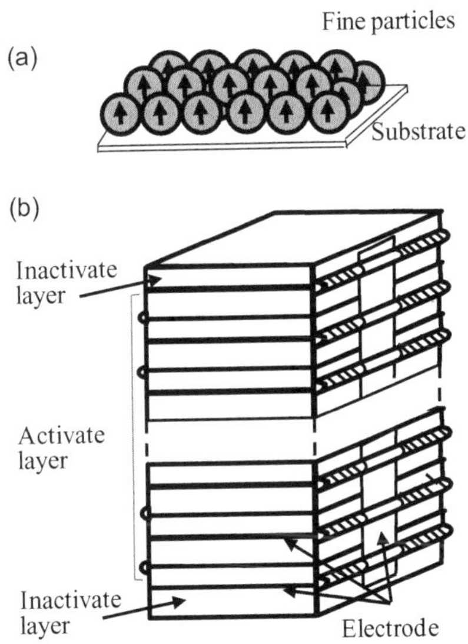

Figure 1. Preferable orientated direction and application for multi layered piezoelectric device.

show superior piezoelectric performance along to a limited crystal axis[23-36]. Bismuth titanate family, for example, which is promising as one of the next generation ferroelectric materials, needs the crystal orientation of ***a***, ***b*** axis for their superior performance[23-30]. Those particles show the plate-like shape with main face of c-face due to crystal anisotropy. This high anisotropy means that a special processing is required to achieve useful properties in the material[19-21]. In application as the piezoelectric materials with sheet-shape, the direction of high spontaneous polarization in the microstructure must be oriented parallel to the electric field. They must be aligned in normal to the substrate with electrode for application use as shown in **Fig.1**. This is very difficult to accomplish with conventional methods such as the hot forging and the doctor blade forming method. The particle orientation assisted by strong magnetic field is very effective for development[10,11].

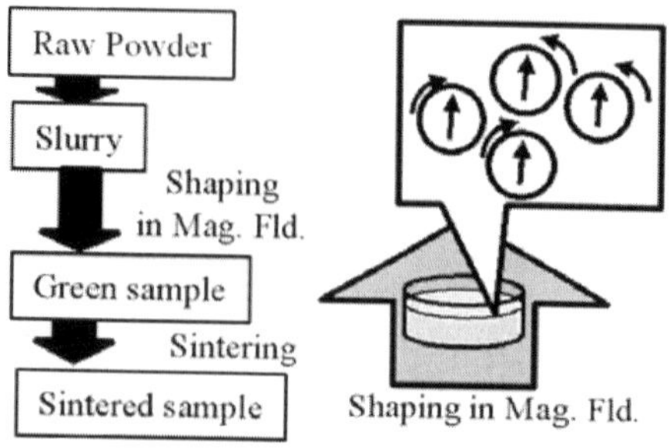

Figure 2. Flow chart of fabrication technique.

The principle of orienting particle in a strong magnetic field is as follows. The interaction of a crystal with a magnetic field is appreciable, even for "non-magnetic" materials such as para- and dia-magnetic materials when the field is very strong, i.e., 10 Tesla by super-conducting magnet. The induced magnetization differs along various principle axes in a non-cubic material placed in a magnetic field. In para- and dia-magnetic materials, the axes of the largest magnetization and the smallest anti-magnetization tend to align along the magnetic field, respectively.

Figure 2 shows the flow chart of fabrication technique (MFSS method). This technique consists from two steps. First step is shaping processing in magnetic field. Second step is sintering processing without magnetic field. Grain growth occurs during sintering. Fine powders always form agglomeration. Those powders are dispersed into water or something liquid using dispersant for free from particle-particle interaction. Then, the slurry in the mold is set into the magnetic field. After dried, the green sample is taken out from the magnetic field. The green sample is sintered at high temperature. The oriented structure is more developed by densification and grain growth.

What are experimental parameters of particle orientation in magnetic field? Two kinds of torques are applied to particle in the slurry which is set in magnetic field. One is the magnetic torque, another is the viscous torque. The magnetic torque is expressed as this equation[37].

$$T = \frac{1}{2}\Delta\chi B^2\left(\frac{4\pi r^3}{3}\right)\sin 2\theta \qquad (1)$$

According to this equation, anisotropy of magnetic susceptibility $\Delta\chi$, particle radius $\boldsymbol{r}$ and magnetic flux density $\boldsymbol{B}$ govern the magnetic torque $\boldsymbol{T}$. On the other hands, particle size and viscosity on particle surface affect the viscous torque. Particularly, viscosity is depended on viscosity of liquid, particle-particle-interaction which is controlled by volume fraction of particles. Of course, the Brownian motion scatters the motion of particle.

What are experimental parameters of texture development during sintering? Densification and grain growth promote the oriented microstructure. Large particles tend to be aligned well by the magnetic torque. Small particles are absorbed in large oriented particles during sintering. Particularly, grain growth occurred well at final stage of sintering. Both of the distribution of primary particle size and the orientation degree in green compact affect this process.

The objective of this study was to fabrication of strontium bismuth titanate ceramics with highly crystal-oriented structure and high density, and to examine the influences of experimental parameters such as magnetic field and sintering on the oriented structure development.

Figure 3. SEM micrographs of the strontium bismuth titanate ($SrBi_4Ti_3O_{12}$, SBTi) powder (a) 0.5, (b) 0.78, (c) 0.8, and (d) 1.0μm.

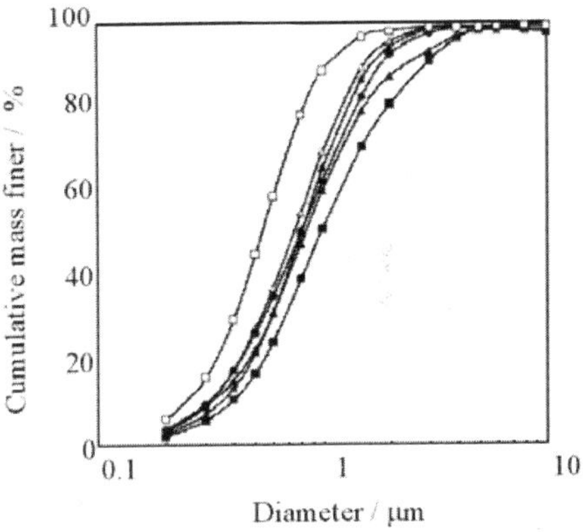

Figure 4. The particle size distributions measured by the sedimentation method using x-ray.

2. EXPERIMENTAL

Strontium bismuth titanate powder ($SrBi_4Ti_4O_{15}$, SBTi) is used as raw materials. SBTi powders were synthesized by solid reaction processing of Bi_2O_3, TiO_2, $SrCO_3$. The synthesized temperatures are 850 and 900°C. The powder was ground for 1-24hours for controlling particle size. The mean sizes of SBTi particles are in the range of 0.5 to 1μm. The crystal phase was characterized by XRD patterns. The diameter distribution was evaluated by X-ray adsorption during sedimentation. Slurry with solid loading 30 vol% was prepared by ball milling for 1 h and poured into a plastic mold set in the magnetic field (10 Tesla). After dried, samples were sintered at 1200 °C for 2h. The orientation was evaluated by XRD. The Lotgering factor F[38] as an index of degree of orientation[39], which was calculated by using the following equation,

$$F = \frac{P - P_0}{1 - P_0} \tag{2}$$

where, $P_0 = \Sigma I_0(hk0)/\Sigma I_0(hkl)$ and $P = \Sigma I(hk0)/\Sigma I(hkl)$. I and I_0 are the intensities of each of the diffraction peaks in XRD patterns as random sample and those determined experimentally, respectively.

3. RESULTS and DISCUSSION

Figure 3 shows SEM micrographs of the strontium bismuth titanate ($SrBi_4Ti_4O_{15}$, SBTi) powders synthesized in this study. The shape of particle was irregular, not equi-axis. Some aggregates were also observed. The larger particles tend to have a platelet shape. The major face of the large plate-like particles should be the *c* face of SBTi. In a single crystal, the *c* axis of the SBTi lies perpendicular to the major face of plate-like particles. **Figure 4** shows the particle size distributions measured by the sedimentation method using x-ray. The particle size had wide distributions. The mean size of particles was in the range of 0.5 to 1.0μm. The synthesized temperature and keeping time increased particle size.

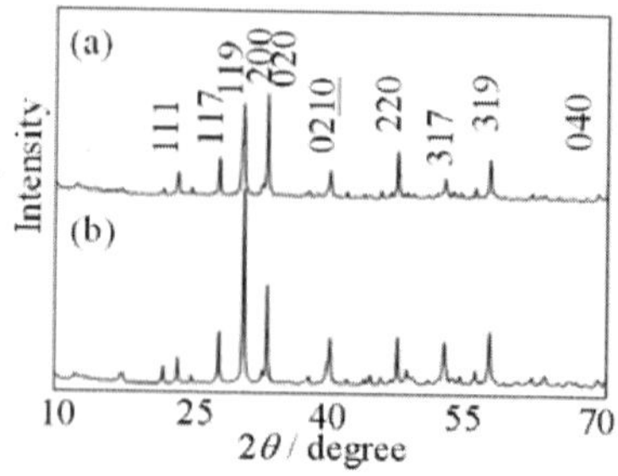

Figure 5. X-ray diffraction patterns of the green specimens (a) in the plane perpendicular to the magnetic field, (b) sample formed without magnetic field.

Figure 5 shows the X-ray diffraction patterns of the green specimens in the plane perpendicular to the magnetic field. The mean diameter 0.8μm SBTi powder was used as a raw material. In the specimen prepared in the high magnetic field, the strong diffraction peaks were those associated with the ***a,b*** planes of the crystal, 200, 020 and 220. **Figure 6** shows the x-ray diffraction patterns of the specimens sintered at 1200 °C. Measured face was the same with that in **Fig.5**. In the specimen shaped in the high magnetic field, the strong diffraction peaks were those associated with the ***a,b*** planes of the crystal. The strongest peak 119 was almost absent from crystal oriented sintered ceramics. The result shows that sintering processing remarkably enhanced the oriented structure.

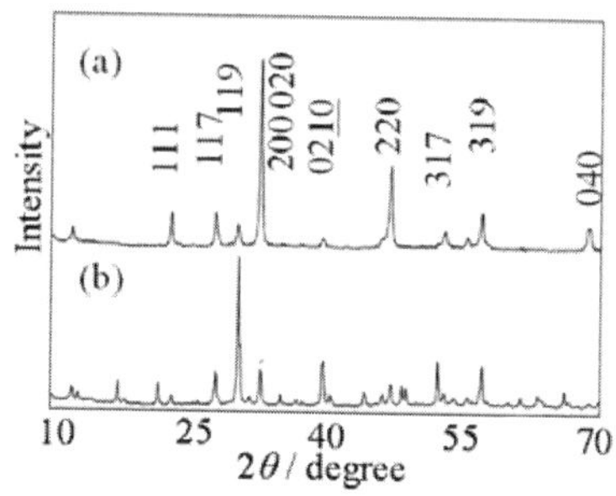

Figure 6 X-ray diffraction patterns of the sintered specimens at 1200 °C (a) in the plane perpendicular to the magnetic field, (b) sample formed without magnetic field.

Figure 7 shows the microstructure of the sintered SBTi in perpendicular and parallel to the magnetic field. Those samples were made from SBTi particles with 0.8μm and heated 1200 °C. The relative density after sintering was high 98% of theoretical density. Both of highly oriented structure and high density was to be expected for the MFSS processing. This result is contribution of fine particles used as raw material. Different microstructures are noted in the two directions, suggesting that the ceramics has an anisotropic structure. The majority of particles appeared to have a plate-like shape, and was randomly oriented in the microstructure observed in the direction parallel to the magnetic field. On the other hand, plate-like particles were observed in the direction perpendicular to the magnetic field. The plate-like grains tended to orient with their longest axis parallel to the direction of the applied magnetic field. The SBTi crystal has a tendency to form a plate shape, with the *c* plane being the largest face of the grain. The grains appeared elongated when viewed from the direction of ***a*** and ***b*** axes. They appeared plate-shaped

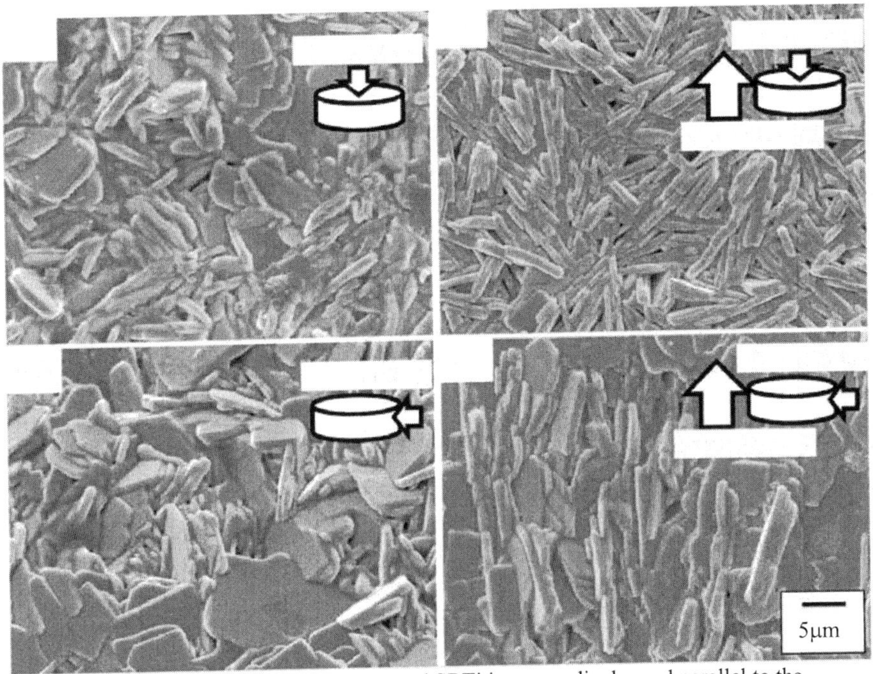

Figure 7. The microstructure of the sintered SBTi in perpendicular and parallel to the magnetic field. (a),(c) samples formed without magnetic field, (b),(d) with magnetic field.(a),(b) cross section in parallel to top surface, (c),(d) perpendicular to top surface.

when viewed from the ***c*** axis. Elongated and plate shape particles were noted as observed in the direction perpendicular to the applied magnetic field, in which the direction of the ***c*** plane was randomly oriented.

Figure 8 shows the Lotgering factors of ***a***, ***b*** planes in both of green samples and sintered samples. The relative densities of the sintered samples are also shown in this figure. The Lotgering factor increased with particle size of SBTi raw powder. They were lower level in all green samples. The particle rotation is inhibited by the interactions between particles, agglomeration, and instability of dispersion, although the magnetic torque ***T*** is proposed to particle volume, cubic of particle radius r^3. After sintering, the Lotgering factor increased well. The

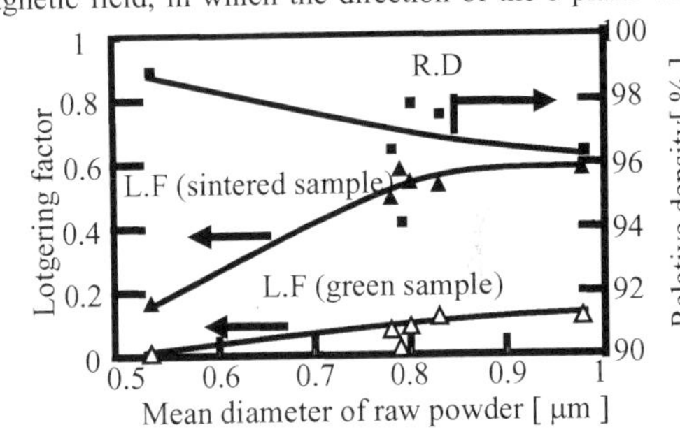

Figure 8. The Lotgering factors of the ***a***, ***b*** planes in both of green samples and sintered samples. The relative densities of sintered samples are also shown in this figure.

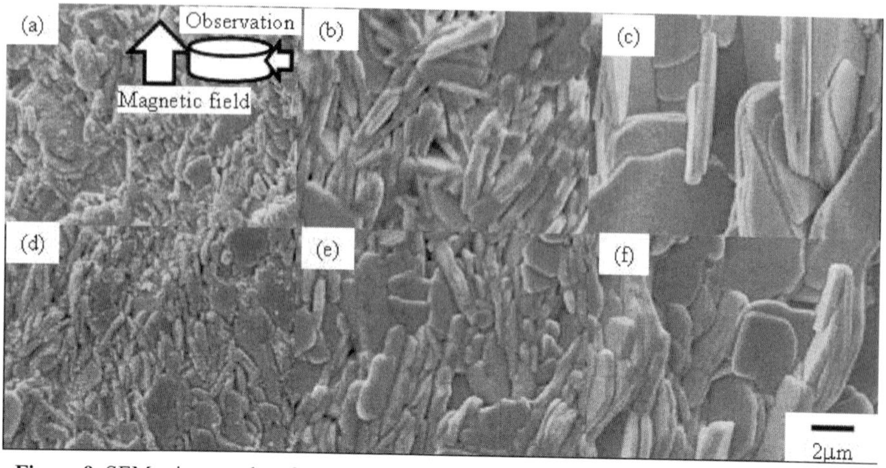

Figure 9. SEM micrographs of microstructure of the sintered samples. The mean sizes of SBTi raw powder were (a)-(c) 0.5 and (d)-(f) 0.8μm. Observed face was the cross section in parallel to magnetic field, (a),(d) 1050°C, (b),(e) 1100°C, (c),(f) 1200 °C.

higher was the Lotgering factor at green samples, the higher was it at sintered samples. The maximum value of Lotgering factor was about 0.6 in the case of SBTi particles with 0.8μm above. On the contrary, the relative density decreased with increasing in mean particle size as known well in ceramic processing. This result shows that the appropriate condition for high dense and high orientation is about 0.8 μm.

Figure 9 shows micrographs of microstructure development of the samples on sintering. The mean sizes of SBTi raw powder were 0.5 and 0.8μm. The particles showed grain growth with sintering increasing for both samples. Although the Lotgering factor of green samples were low level 0.08 and 0.2, the difference between sintered samples heated at 1200°C were considerably larger. When the size of raw powder was 0.5μm, the both of oriented and not-oriented particles became larger with maintaining each direction. This slight difference of oriented structure in green compacts between both samples governs the microstructure of sintered samples.

CONCLUSIONS

Fabrication of strontium bismuth titanate ceramics $SrBi_4Ti_4O_{15}$ with highly crystal-oriented structure and high density was examined for the magnetic field-assisted shaping followed by conventional sintering processing (MFSS) method. The influence of particle size on the orientation structure was important for each fabrication step; shaping in magnetic field and sintering processing. The Lotgering factor as an index of orientation degree of the green sample shaped in strong magnetic field 10 Tesla was slightly increased from 0.08 to 0.2 with particle size 0.5-1.0μm. After sintering, it increased to 0.2-0.6. The remarkable development of orientation structure during sintering was observed by SEM. On the contrary, the relative density was decreased with particle size. When using large particle size 0.8μm, the high orientation degree and high relative density were achieved.

REFERENCE

[1] P.De,Rango, M.Lees, P.Lejay, A.Sulpice, R.Tournier, M.Ingold, P.Germi and M.Pernet, Texturing of magnetic materials at high temperature by solidification in a magnetic field, *Nature*, 349, 770-71 (1991).

[2] J.G.Noudema, J.Beilleb, D.Bourgaulta, D.Chateignerc and R.Tourniera, Bulk textured BiPbSrCaCuO (2223) ceramics by solidification in a magnetic field, *Physica C*, 264, 325-30 (1996)

[3] Y.Nakagawa, H.Yamasaki, H.Obara and Y.Kimura, Superconductiong properties of grain-oriented samples of $YBa_2Cu_3O_y$, Jpn,J.Appl. Phys. 28, 4, L547-50 (1989)

[4] M. H. Zimmerman, K. T. Faber, E. R. Fuller Jr., K. L. Kruger and K. J. Bowman, Texture assesment of magnetically processed iron titanate, *J.Am.Ceram.Soc.*, **79**, 1389-93 (1996)

[5] M. H. Zimmerman, K. T. Faber and E. R. Fuller Jr, Forming textured microstructures via the gelcasting technique, *J.Am.Ceram.Soc.*, **80**, 2725-29 (1997)

[6] K.Uematsu, T.Ishikawa, D.Shoji, T.Kimura, K.Kitazawa, Japan-Patent 3556886

[7] T.S.Suzuki, Y.Sakka and K.Kitazawa, Orientation amplification of alumina by colloidal filtration in a strong magnetic field and sintering, *Adv. Engineering Mater.*, **3**, 44-46 (2001)

[8] A.Makiya, D.Shoji, S.Tanaka, N.Uchida, T.Kimura and K.Uematsu, Grain Oriented Microstructure Made in High Magnetic Field, *Key Engineering Mater.*, **206-213**, 445-48(2002).

[9] A.Makiya, K.Kusumi, S.Tanaka, K.Uematsu, Particle oriented titana ceramics prepared in a magnetic field, *J.Euro.Ceram.Soc.*, **27**, 797-99 (2007)

[10] A.Makiya, D.Kusano, S.Tanaka, N.Uchida, K.Uematsu, T.Kimura, K.Kitazawa and Y.Doshida, Particle oriented bismuth titanate ceramics made in high magnetic field, *J. Ceram. Soc. Japan*, **111**, 702-4 (2003)

[11] Y.Doshida, K.Tsuzuku, H.Kishi, A.Makiya, S.Tanaka, K.Uematsu and T.Kimura, Crystal-oriented $Bi_4Ti_3O_{12}$ ceramics fabricated by high-magnetic-field method, *J. J. Appl. Phys.*, **43**, 6645-48 (2004)

[12] T.S.Suzuki and Y.Sakka, Control of texture in ZnO by slip casting in a strong magnetic field and heating, *Chem. Lett.*, 1204-5 (2002)

[13] T.S.Suzuki and Y.Sakka, Fabrication of textured titania by slip casting in a high magnetic field followed by heating, *Jpn J. Appl. Phys.*, **41,** L1272-74 (2002)

[14] Y.Sakka and T.S.Suzuki, Textured development of feeble magnetic ceramics by colloidal processing under high magnetic field, *J.Ceram.Soc.Jpn.*, **113,** 26-36 (2005)

[15] S.Tanaka, A.Makiya, Z.Kato, N.Uchida, T.Kimura and K.Uematsu, Fabrication of c-axis orientated polycrystalline ZnO using a rotating magnetic field and following sintering, *J.Material Res.*, **21,** 703-7 (2006)

[16] Y.Doshida, H.Kishi, A.Makiya, S.Tanaka, K.Uematsu and T.Kimura, Crystal-oriented La-substituted $Sr_2NaNb_5O_{15}$ ceramics fabricated using high-magnetic-field method, *J. J. Appl. Phys.*, **45**, 7460-64 (2006)

[17] S.Tanaka, A.Makiya, T.Okada, T.Kawase, Z.Kato K.Uematsu, C-axis orientation of $KSr_2Nb_5O_{15}$ by using a rotating magnetic field, *J.Am.Ceram.Soc*, **90**, 3503-06 (2007)

[18] Y.Saito, H.Takao, T.Tani, T.Nonoyama, K.Takatri, T.Homma, T.Nagaya and M.Nakamuram, Lead-free piezoceramics, *Nature*, **432**, 84-87 (2004)

[19]G.L.Messing, S.Trolier-McKinstry, E.M.Sabolsky, C.Duran, S.Kwon, B.Brahmaroutu, P.Park, H.Yilmaz, P.W.Rehrig, K.B.Eitel, E.Suvaci, M.Seabaugh, and K.S.Oh, Templated grain growth of textured piezoelectric ceramics, *Solid State and Materials Sciences*, **29**, 45-69 (2004)
[20]T. Kimura, Application of textured engineering to piezoelectric ceramics, *J. Ceram. Soc. Jpn*, **114**, 15-25 (2006)
[21]T. Tani, Texture engineering of electronic ceramics by the reactive-template grain growth method, *J. Ceram. Soc. Japan*, **114**, 363-70 (2006)
[22]T. Takenaka, Grain Orientation Effects on Electrical Properties of Bismuth Layer-structured Ferroelectric Ceramics, *J. Ceram. Soc. Japan*, **110**, 215-24 (2002).
[23]S. Ikegami and I. Ueda, Piezoelectricity in Ceramics of Ferroelectric Bismuth Compound with Layer Structure, *Jpn. J. Appl. Phys.*, **13**, 1572-77(1974).
[24]T. Takenaka and K. Sakata, Grain Orientation and Electrical Properties of Hot-forged $Bi_4Ti_3O_{12}$ ceramics, *Jpn.J. Appl. Phys.*, **19**, 31-39(1980).
[25]S.E.Cummins and L.E.Cross, Electrical and Optical Properties of Ferroelectric $B_{i4}Ti_3O_{12}$Crystals, *J. Appl. Phys.* **39,** 2268-74(1968),
[26]Y. Inoue, T. Kimura and T. Yamaguchi, Sintering of Plate-Like $B_{i4}Ti_3O_{12}$ Powders," *Am. Ceram. Soc. Bull.*, **62,** 704-7,711(1983).
[27]H. Watanaba, T. Kimura and T. Yamaguchi, Sintering of Platelike Bismuth Titanate Powder Compacts with Preferred Orientatio", *J. Am. Ceram. Soc.*, **74**, 139-47(1991).
[28]M. M. Seabaugh, I. H. Kerscht and G.. L. Messing, Texture Development by Templated Grain Growth in Liquid-Phase-Sintered α-Alumina, *J.Am.Ceram.Soc.*, **80,** 1181-88(1997).
[29]T.Kimura, S.Miyamoto and T.Yamaguchi, Microstructure development and dielectric propertie s of potassium strontium niobate ceramics, *J. Am. Ceram. Soc.*, **73**, 127-30 (1990)
[30]Z.Chen, A.Shui, Z.Lu and P.Liu, Piezoelectric and dielectric properties of $(Bi_{0.5}Na_{0.5})TiO_2$-Ba($Zr_{0.04}Ti_{0.96})O_3$ lead-free piezoelectric ceramics, *J. Ceram. Soc. Japan*, **114**, 857-60(2006)
[31]R. R. Neurgaonkar, W. W. Ho, W. K. Cory, W. F. Hall, and L .E. Cross, Low and high frequen cy dielectric properties of ferroelectric tungsten bronze $Sr_2KNb_5O_{15}$ crystals, *Ferroelectrics*, **51**, 185-91 (1984)
[32]K. Matsuo, R. J. Xie, Y. Akimune, and T. Sugiyama, Preparation of lead free $Sr_{2-x}Ca_xNaNb_5O_{15}$ based piezoceramics with tungsten bronze structure, *J. Ceram. Soc. Japan*, **110**, 491-4 (2002)
[33]R.R.Neurgaonkar, J.R.Oliver, W.K.Cory, L.E.Cross and D.Viehland, Piezoelectricity in tungste n bronze crystals, *Ferroelectrics*, **160**, 265-6 (1994)
[34]M.Granahan, M.Holmes, W.A.Schulze, and R.E.Newnham, Grain-oriented $PbNb_2O_6$ ceramics, *J. Am. Ceram. Soc.*, **64**, C68-9(1981)
[35]C.Duran, S.Trolier-McKinstry and G..L.Messing, Fabrication and Electrical properties of textur ed $Sr_{0.47}Ba_{0.47}Nb_2O_6$ ceramics by templated grain growth, *J. Am. Ceram. Soc.,* **83**, 2203-13(2000)
[36]C.Duran, S.Trolier-McKinstry, and G.L.Messing, Dielectric and piezoelectric properties of textured $Sr_{0.47}Ba_{0.47}Nb_2O_6$ ceramics by templated grain growth, *J. Mater. Res.*, **17**, 2399-2409(2002)
[37]T.Kimura, Study on the effect of magnetic fields on polymeric materials and its application, *Polymer J.*, **35**, 823-43 (2003)
[38]F.K.Lotgering, Topotactial reactions with ferromagnetic oxides having hexagonal crystal struct ure, *J. Inorg. Nucl. Chem.*, **9,** 113-23 (1959)
[39]J.L.Jones, E.B.Slamovich and K.J.Bowman, Critical evaluation of the Lotgering degree of orientation texture indicator", *J. Mater. Res.*, **19**, 3414-22 (2004)

RECENT INVESTIGATIONS OF Sr-Ca-Co-O THERMOELECTRIC MATERIALS

W. Wong-Ng, G. Liu, M. Otani, E. L. Thomas, N. Lowhorn, M.L. Green
Materials Science and Engineering Laboratory
National Institute of Standards and Technology
Gaithersburg, MD 20899

J.A. Kaduk
INEOS Technologies
Naperville, IL 60566

ABSTRACT

Three low-dimension cobaltites in the Sr-Ca-Co-O system have been studied for their structure and thermoelectric properties. Using x-ray pole figure construction technique, a $Ca_3Co_4O_9$ thin film showed excellent fiber texture but no *ab* in-plane texture. The fiber texture is mainly responsible for the excellent thermoelectric properties. Our study of a high-throughput combinatorial thin film (synthesized using Pulsed Laser Deposition) with a custom-built power factor screening tool demonstrated that the La substitution in the Ca-site of $Ca_3Co_4O_9$ increases Seebeck coefficient and resistivity due to decrease of hole concentration. The Sr-substitution on the Ca-site increases the carrier mobility, therefore lower resistivity of the film. The homologous compounds $Ca_{n+2}Co_nCo'O_{3n+3}$ and $Sr_{n+2}Co_nCo'O_{3n+3}$ consist of parallel 1-dimensional cobalt oxide chains that are built by successive alternating face-sharing CoO_6 trigonal prism and 'n' units of CoO_6 octahedra along the *c*-axis. However, due to the high resistivity of these compounds, they will unlikely yield *ZT* values comparable to that of $Ca_3Co_4O_9$.

INTRODUCTION

In the last decade, increasing global interest in research and development of thermoelectric materials was partly due to the soaring energy demand and partly due to the need to create a sustainable future. The efficiency and performance of thermoelectric energy conversion or cooling is related to the dimensionless figure of merit (*ZT*) of the thermoelectric materials, given by

$$ZT = \frac{S^2 \sigma T}{k}, \quad (1)$$

where T is the absolute temperature, S is the Seebeck coefficient [1] or thermoelectric power, σ is the electrical conductivity, and k is the thermal conductivity. ZT is directly related to the coefficient of performance of a thermoelectric material and is the reference by which these materials are judged. Thermoelectric materials with desirable properties (high ZT (>>1), i.e., characterized by high electrical conductivity, high Seebeck coefficient and low thermal conductivity) will have widespread military and industrial applications.

For the thermoelectric technology to hold promise for large-scale applications, efforts to identify improved materials and optimize existing materials are crucial. For example, materials that are stable at high temperature would be important for applications in automobile industry.

For almost half a decade, only a small number of materials have been found to have practical industrial applications and they all have *ZT* values around or below 1.0. Recent reports that relatively high *ZT* values are possible in both thin films and bulk forms [2, 3, 4] have revitalized new materials development for thermoelectric industry. For example, Venkatasubramanian et al. found a *ZT* of about 2.4 at 300 K in Bi_2Te_3/Sb_2Te_3 p-type superlattices, and 1.4 in $Bi_2Te_3/Bi_2Te_{2.83}Se_{0.17}$ n-type superlattices [5]. Other state-of-the-art materials include quantum well films [6] and quantum dot films [7] that yield *ZT* as high as 3.0.

The discovery of improved thermoelectric oxides has been of great interest for high temperature thermoelectric applications because of the stability of oxides at high temperature. The low dimension cobaltites that include $NaCoO_x$ [8], $Ca_2Co_3O_6$ [9, 10], and $Ca_3Co_4O_9$ [11-13] have attracted considerable attention because of the coexistence of large Seebeck coefficient and relatively low thermal conductivity. For example, $Ca_3Co_4O_9$ is a misfit layered oxide that has two monoclinic subsystems with identical *a*, *c*, *β*, but different *b* [12]. The 1st subsystem consists of triple rock-salt layers of Ca_2CoO_3 in the *ab* plane, and the second subsystem consists of single CoO_2 layer, which is of the CdI_2-type structure. This phase exhibits strong anisotropic thermoelectric properties in *ab*-plane. It is thought that the increased scattering of phonons at the interface of misfit layers led to the lowering of the lattice thermal conductivity.

Because of the promising properties of the calcium-containing cobaltite, the strontium-doped calcium cobaltites may also offer desirable properties. This paper summarizes our recent investigation of selected series of compounds in the Sr-Ca-Co-O system, including property optimization of $Ca_3Co_4O_9$ using the combinatorial compositional spread approach, texture characterization of a $Ca_3Co_4O_9$ film, and structure/property relation study of a homologous series, $(Sr, Ca)_{n+2}Co_nCo'O_{3n+3}$.

EXPERIMENTAL[1]

(i) Polycrystalline Sr-Ca-Co-O Sample Preparation

Polycrystalline samples in the Sr-Ca-Co-O system were prepared using the conventional high-temperature solid-state synthesis technique from commercial $CaCO_3$ (99.99 %), $SrCO_3$ (99.9 %), and Co_3O_4 (99.99 %). Powders were weighed in stoichiometric proportions, mixed and calcined at 750 °C overnight in air to assure carbonate decomposition, and then reground and reheated at 850 °C for 2 weeks with intermediate grindings and pelletizations. The annealing process was repeated until no further changes were detected in the powder X-ray diffraction patterns. Phase analyses were carried out using a Phillips X-ray powder diffractometer in θ-2θ scan mode with Cu Kα radiation and equipped with a series of Soller slits and a scintillation counter. The 2θ scanning range was from 12° to 81°, and the step interval was 0.03°.

(ii) $Ca_3Co_4O_9$ Texture Determination

(a) <u>Thin Film preparation</u>

Details of the $Ca_3Co_4O_9$ thin film preparation have been reported elsewhere [14]. In brief, the film was deposited on a commercial Si (100) single crystal wafer in an *in situ* manner using the Pulsed Laser Deposition (PLD) facility at the Brookhaven National Laboratory. The $Ca_3Co_4O_9$

[1] The purpose of identifying the equipment and computer software in this article is to specify the experimental procedure. Such identification does not imply recommendation or endorsement by the National Institute of Standards and Technology.

target was prepared from a stoichiometric mixture of high-purity $CaCO_3$ and Co_3O_4 powders. The powders were homogenized and heat-treated twice at 880 °C to 890 °C for 24 h in flowing air with intermediate grindings. After pressing the powder into a disk, the final sintering was accomplished at 900 °C for 24 h in flowing O_2 gas. A film about 2300 Å thick was deposited at a substrate temperature of 700 °C with a laser energy density of $\approx$ 1.5 J/cm^2, under an oxygen partial pressure of 39.5 Pa (or 300 mTorr). After deposition, films were cooled to room temperature in $\approx$ 0.1 MPa of oxygen.

(b) Texture Characterization

Data collection for film texture was performed on a Bruker D8 Diffractometer with the General Area Detector Diffraction System (GADDS) and a rotating anode, operating at 40 kV and 20 mA. The core of the GADDS is the two-dimensional area detector that is a photon counter over a large area [15]. The goniometer system is oriented horizontally with the sample mounted on a χ, ϕ eucentric cradle so that the film is normal to the (azimuthal) ϕ axis. One can obtain both qualitative and quantitative texture information by evaluating the Debye rings. For example, if the Debye rings are continuous and smooth, the sample is polycrystalline and fine grained. If the rings are continuous but spotty, the material is polycrystalline and contains large grains. Incomplete Debye rings indicate orientation or texture. If only individual spots are observed, the material is a single crystal or the extreme case of crystallographic texture.

Pole figure constructions were performed using the software system that is part of GADDS [15]. GADDS was used to measure pole figures at fine steps of 2° in ϕ that allow for detection of sharp texture. Data collection for texture analysis have been performed on the $Ca_3Co_4O_9$ film using the 005 reflection for determining the (00ℓ) texture, and using the 20 -1, 11-2, 111, 203, 202, and 020 reflections for in-plane texture study. The 2D patterns were recorded with χ values from 10° to 80° and 2θ from 25° to 50°. Once the pole figure frames were collected, integrations of the reflections of interest in each of the frames were carried out. At each of the χ angles 10°, 35°, and 60°, 180 frames were collected; and at χ=80°, 120 frames were collected.

(iii) Combinatorial Film preparation

The (Ca, Sr, La)$_3$Co$_4$O$_9$ composition-spread films were fabricated with a combinatorial pulsed laser deposition system by a continuous-composition-spread technique [16, 17]. In summary, a $Ca_3Co_4O_9$ film was deposited first by ablation of a $Ca_3Co_4O_9$ target by a KrF excimer laser. The resulting film had a thickness distribution, the center of which is away from that of the substrate. The thickness of the 'thickness center' corresponds with to one monolayer of $Ca_3Co_4O_9$. The substrate was then rotated 120° and a $(Ca_2Sr)Co_4O_9$ film was deposited. Another rotation of 120° and a $(Ca_2La)Co_4O_9$ film was further deposited. This process was repeated a few hundred times to obtain the desired thickness for thermoelectric measurements. The film was grown at a growth temperature of 750 °C and p_{O2} of 80 Pa on a 76.2 mm Si (100) wafer.

To estimate the composition and the total thickness of the composition-spread film, reference single layer films of the constituents of the compositional spread films $Ca_3Co_4O_9$, $(Ca_2Sr)Co_4O_9$, and $(Ca_2La)Co_4O_9$ for the $(Ca_{1-x-y}Sr_xLa_y)_3Co_4O_9$ film were fabricated under the same conditions as those of the composition-spread films. Since the electric resistance of a thin film is proportional to its thickness, the relative thickness for each point of a single layer film can be evaluated by the four-probe method with the screening tool. The thickness of the 'thickness center' was measured by a wet-etching technique and by an atomic force microscope. From the thickness distributions of all single layer films, we estimated the composition and the total thickness of the composition-spread film.

(iv) Thermoelectric Property Measurements

(a) Thin films

We have developed a high-throughput power factor ($S^2\sigma$) screening tool system (Fig. 1) that can be used to measure electrical conductivity and Seebeck coefficient [16, 17] of thin films. The central feature of the screening tool consists of a measurement probe to measure electric conductivity and Seebeck coefficient, an automated translation stage to scan the probe in the *x-y-z* directions, and various electrical measuring instruments. The measurement probe consists of four spring probes as sample contacts, a heater to generate temperature differences between two of the spring probes, and two thermometers to measure the temperature of these probes. Measurements are fully automated by a laptop computer. The measurement probe and a sample stage are placed inside a protective case to stabilize the temperature of the measurement probe. Electric conductivity is measured by the conventional 4-probe method. The Seebeck coefficient is defined as the ratio of voltage (ΔV) and temperature ΔT at two points. It takes 20 seconds to measure both electric conductivity and Seebeck coefficient for each sample point. This probe allows us to measure electric conductivity and Seebeck coefficient of over 1000 sample points within 6 hours. All Seebeck coefficient measurements were conducted at room temperature and at $\Delta T = 4.1$K.

(b) Bulk Samples

The bulk samples were pressed into bars (approx. dimensions of 3.5x2.5x6 mm^3) and sintered at 850 °C. The high throughput screening tool as mentioned before [16, 17] was first used to screen the Seebeck coefficient and resistivity of the flat surface of the bulk samples to ensure a reasonable power factor value before the use of the relatively time consuming bulk technique. A Physical Property Measurement System (PPMS) was used to measure the final Seebeck coefficient, resistivity, and thermal conductivity. The samples were measured from 4 K to 390 K.

RESULTS AND DISCUSSION

Three phases were found in the Sr-Ca-Co-O system. The first one is the $(Ca, Sr)_3Co_4O_9$ solid solution that features the misfit layered structure. The second phase, $(Ca,Sr)_3Co_2O_6$, has linear $Co_2O_6^{6-}$ chains. The third phase is a homologous series of compounds, $A_{n+2}B_nB'O_{3n+3}$ (A=Sr,Ca; B=B'=Co), that also possess long cobalt oxide chains. In the following, we will summarize our studies of the texture and property optimization of $Ca_3Co_4O_9$, and the structure/ thermoelectric property characterization of $(Sr,Ca)_{n+2}Co_nCoO_{3n+3}$.

(i) *Texture of $Ca_3Co_4O_9$ thin film*

The (00ℓ) texture of the $Ca_3Co_4O_9$ thin film was characterized using an area detector frame in which the 005 arc was clearly visible [18]. Integration of the 005 arc gave a texture profile that was perfectly symmetrical about its maximum angle. The two parts of the texture profile on either side of the maximum angle ($\chi < 90°$ and $\chi > 90°$) were averaged; and the resulting intensity plot, $I(\chi)$, is shown in Fig. 2. The volume fraction, $V(\alpha)$, of the film with its (00ℓ) pole within angle α of the film normal is given by:

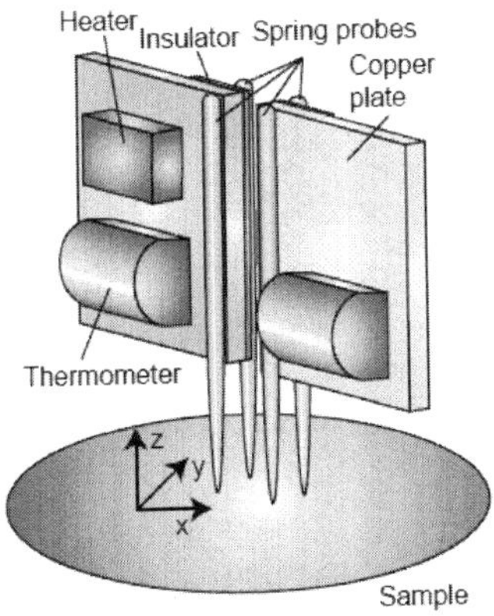

Fig. 1. Power factor screening tool for combinatorial thin film [16, 17].

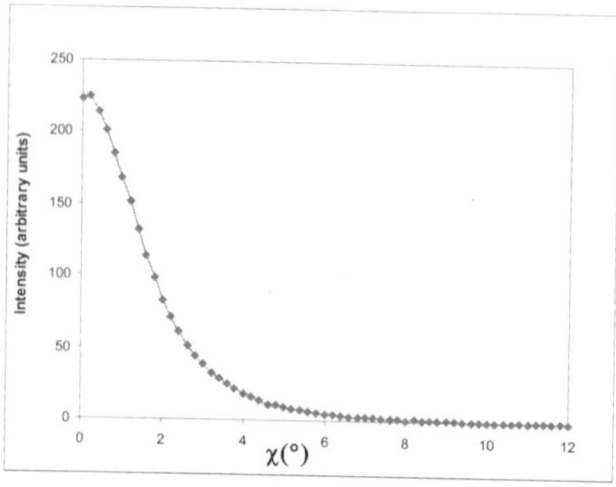

Fig. 2. Averaged texture profile, I(χ), of the 005 reflection [14].

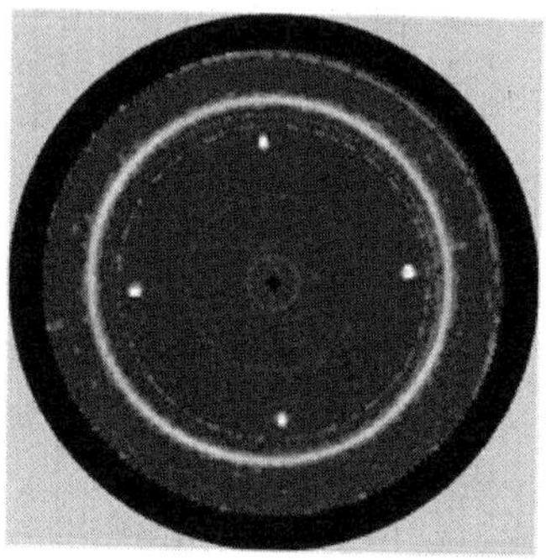

Fig. 3. (111) pole figure (continuous ring) of the $Ca_3Co_4O_9$ film deposited on a single crystal Si substrate (2θ range from 28.4° to 29.4°). The diffraction spots are due to the overlap with the (111) reflection of the Si substrate. A series of rings with very weak intensity are an artifact of data analysis [14].

$$V(\alpha) = \int_0^{\alpha} I(\chi)\sin\chi d\chi \Big/ \int_0^{90} I(\chi)\sin\chi d\chi \qquad (2)$$

where it is assumed that the intensity is zero from 12° to 90° [19]. For example, we found that two thirds of the film is oriented with (00ℓ) within 3° of the film normal and 94 % of the film within 6°. Therefore the sample has an excellent (00ℓ) texture.

There is no evidence for the in-plane (*ab* plane) texture for the $Ca_3Co_4O_9$ film. The pole figure obtained using reflection such as 111 (Fig. 3) indicated the absence of an in-plane epitaxial relationship between the film and the substrate. Positions of features on pole figures are characterized by their polar angle, α, and azimuthal angle, β. In these pole figures, strong peaks that represent single crystal-like texture are absent, while continuous rings of intensity are observed, showing random in plane orientation of the grains in the film; the ring in Fig. 3 is at $\alpha = 68.0°$. These compare very well with the calculated angles in the $Ca_3Co_4O_9$ structure: 67.9° between (001) and (111). It is therefore concluded that the film does not have *ab*-plane texture but has a strong (00ℓ) fiber texture. The diffraction spots in Fig. 3 are due to the tail of the strong (111) reflection of the Si substrate.

The absence of *ab*-plane alignment of the $Ca_3Co_4O_9$ film with the substrate does not appear to be an important factor that contributes to the overall property. Strong (00ℓ) fiber texture, on the other hand, is critical for achieving excellent thermoelectric properties for these films [20].

(2) *Combinatorial Study of (Ca,Sr,La)$_3$Co$_4$O$_9$ Films*

In Fig. 4 the power factor ($S^2\sigma$) of the $(Ca_{1-x-y}Sr_xLa_y)_3Co_4O_9$ ternary composition-spread film is depicted on a conventional ternary diagram measured with the screening tool [16, 17]. It is clear that the power factor peaks between the Sr-rich region and the La-rich region. This result is reasonable since electric conductivity is high in the Sr-rich region, and Seebeck coefficient is alarge in the La- rich region. As the La substitution for Ca increases, the electric conductivity decreases, and Seebeck coefficient increases. Substitution of the trivalent La^{3+} for the divalent Ca^{2+} is expected to decrease the hole concentration, and therefore the electric conductivity. On the other hand, substitution of Sr for Ca causes increase of electric conductivity and a insignificant change of Seebeck coefficient. This is because the substitution of a larger divalent Sr^{2+} cation for a smaller divalent Ca^{2+} cation would not change the carrier concentration but the carrier mobility by lattice deformation.

(3) *Sructure and thermoelectric properties of $A_{n+2}B_nB'O_{3n+3}$ (A=Sr,Ca; B=B'=Co)*

The homologous series, $A_{n+2}B_nB'O_{3n+3}$ (A=Sr, Ca; B=B'=Co), which consists of 1-dimensional parallel cobalt oxide chains, is built by successive alternating face-sharing CoO_6 trigonal prisms and CoO_6 octahedra along the *c*-axis [21]. This face-sharing feature is in contrast with $Ca_3Co_4O_9$ and $NaCo_2O_4$ which consists of edge-sharing CoO_6 octahedra. In the formula $A_{n+2}B_nB'O_{3n+3}$, A is an alkali-earth element, B describes cobalt inside the octahedral cage, and B' is the cobalt inside a trigonal prism. The cobalt oxide chains can be considered as stack up of alternate 'n' number of octahedra with one trigonal prism.

The compounds of $A_{n+2}Co_nCo'O_{3n+3}$ can be considered as ordered intergrowth between the n=infinitive ($ACoO_3$) and n=1 ($A_3Co_2O_6$) end members. We found that when A=Ca, only the n=1 member, namely $Ca_3Co_2O_6$, can be made. The linear $Co_2O_6^{6-}$ chains of $Ca_3Co_2O_6$, (*R-3c*, *a* = 9.0793(7) Å , and *c* = 10.381(1) Å [22], Fig. 5), consist of one CoO_6 octahedron alternating with one CoO_6 trigonal prism. Each chain is surrounded by the alkali earth elements and six other

$Co_2O_6^{6-}$ chains. Sr is found to substitute in the Ca-site to the extent of $(Ca_{0.9}Sr_{0.1})_3Co_2O_6$ at 850 °C. The Seebeck coefficient for $Ca_3Co_2O_6$ has been found to be relatively high and positive, and the thermal conductivity is relatively low at high temperature. The transport property is dominated mainly by p-type carriers [23]. *ZT* was determined to be about 0.15 at 1000K.

We found that the stable compounds in the $Sr_{n+2}Co_nCo'O_{3n+3}$ series with larger alkaline-earth elements [24] are those with $2 \leq n < 5$. For example, the x-ray diffraction pattern of the n=1 member, $Sr_3Co_2O_6$, gives a mixture of $Sr_4Co_3O_9$ and SrO. The n=2 member, $Sr_4Co_3O_9$, has been reported to be isostructural with $Sr_4Ni_3O_9$ (Fig. 6). The space group and unit cell parameters of $Sr_4Co_3O_9$ are determined to be *P321* and a = 9.5074 Å, and c = 7.9175Å from x-ray diffraction. In the $Co_3O_9^{8-}$ chains, there are two units of CoO_6 octahedra alternating with 1 CoO_6 prism. The structure for the n=3 and n=4 members, namely, $Sr_5Co_4O_{12}$ and $Sr_6Co_5O_{15}$ feature 3 and 4 octahedra interleaving with one trigonal prism along the *c*-axis, respectively. Ca was found to substitute into the Sr sites of these compounds. We are in the process of determining the range of substitution. In summary, three members of the series of $Sr_{n+2}Co_nCoO_{3n+3}$ exist in the Sr-Co-O system, with n=2, 3, and 4.

The Seebeck and resistivity data of selected n=2 and n=4 members ($(Sr_{0.7}Ca_{0.3})_4Co_3O_9$ and $(Sr_{0.87}Ca_{0.13})_6Co_5O_{15}$) as a function of temperature, measured using the PPMS, are given in Fig. 7. While the Seebeck coefficient is in general high for the $A_{n+2}B_nB'O_{3n+3}$ family, however, the resistivity values of most of these samples are relatively high. Unless one can reduce their resisitivity, they will not likely to have practical energy conversion applications.

SUMMARY

We have studied the crystal chemistry, structure, texture, and thermoelectric properties of selected compounds in the Sr-Ca-Co-O system. The Sr-Ca-Co-O system is an important system because it consists of three low dimension phases (misfit layered $Ca_3Co_4O_9$, and $Ca_3Co_2O_6$ and $Sr_{n+2}Co_nCo'O_{3n+3}$ with one dimensional chains) that are of interest to the thermoelectric research field. However, the resistivity values of the samples with one dimensional chains are relatively high. Therefore unless we can decrease the resisitivity via substitution or processing, they are not likely to be considered as potential candidates for energy conversion. At the present time, the best oxide materials for thermoelectric applications are the cobaltites that feature the misfit layered structure, based on $Ca_3Co_4O_9$.

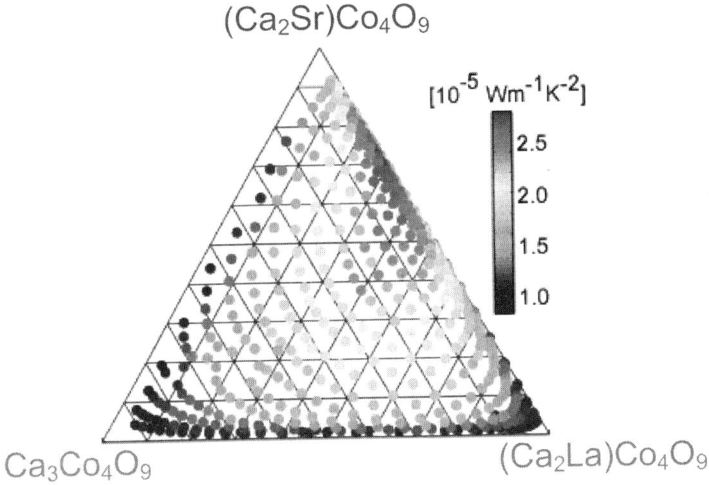

Fig. 4. Power factor diagram of the combinatorial film system of $Ca_3Co_4O_9$- $(Ca_2Sr)Co_4O_9$-$(Ca_2La)Co_4O_9$

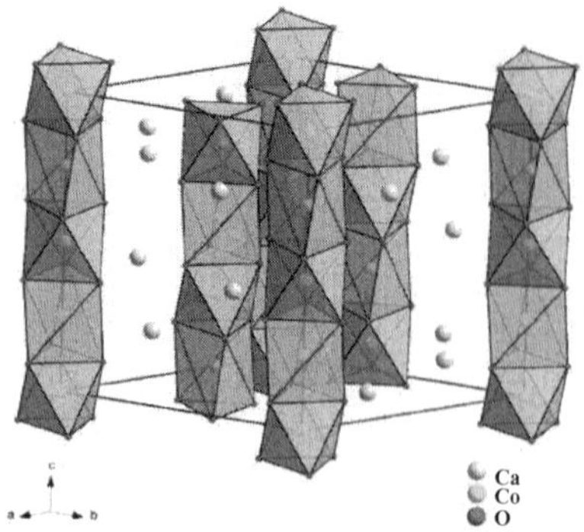

Fig. 5. Crystal structure of $Ca_3Co_2O_6$, n=1 member in the homologous series, $Ca_{n+2}Co_nCoO_{3n+3}$

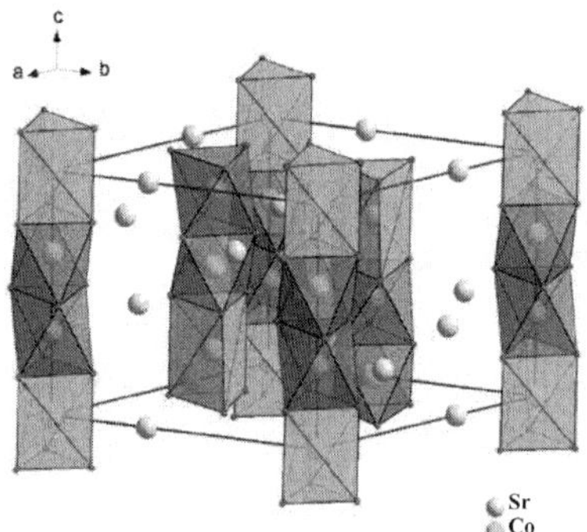

Fig. 6. Crystal structure of $Sr_4Co_3O_9$, n=2 member in the homologous series, $Sr_{n+2}Co_nCoO_{3n+3}$

(a)

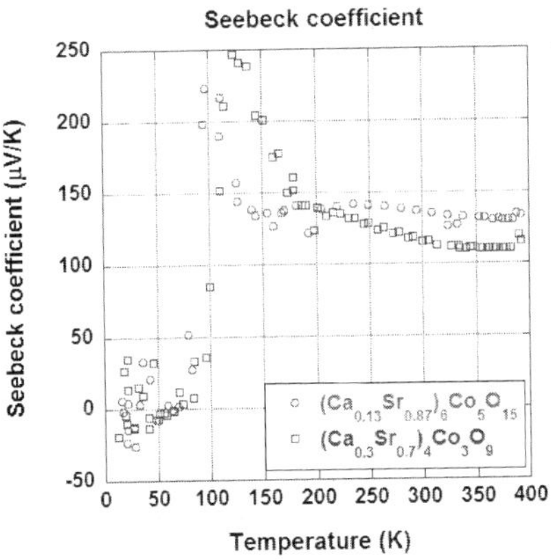

(b)

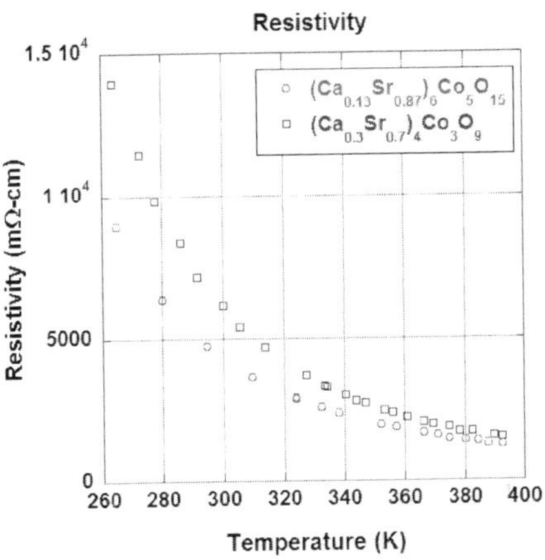

Fig. 7. (a) Seebeck coefficient and (b) resistivity for the n=2 ($(Sr_{0.7}Ca_{0.3})_4Co_3O_9$) and n=4 ($(Sr_{0.87}Ca_{0.13})_6Co_5O_{15}$) members of $(Sr,Ca)_{n+2}Co_nCoO_{3n+3}$

REFERENCES

[1] T.J. Seebeck, *Abhand Deut. Akad. Wiss. Berlin*, 265-373 (1822).

[2] T. M. Tritt and M.A. Subramanian, guest editors, *Harvesting Energy through thermoelectrics: Power Generation and Cooling*, pp. 188-195, MRS Bulletin, Materials Research Soc. (2006).

[3] T. M. Tritt, *Science* **272**, 1276 (1996).

[4] Kuei Fang Hsu, Sim Loo, Fu Guo, Wei Chen, Jeffrey S. Dyck, Ctirad Uher, Tim Hogan, E. K. Polychroniadis, and Mercouri G. Kanatzidis, *Science* **303**, 818 (2004).

[5] R. Venkatasubramanian, E. Siivola, T. Colpitts, and B. O'Quinn, *Nature* **413**, 597(2001).

[6] S. Ghamaty and N.B. Eisner, Proceeding of Interpack 2005: ASME Technical Conference on Packaging of MEMS, NEWS and Electric Systems, July 17-22, San Francisco, CA.

[7] M.S. Dresselhaus, G. Chen, M.Y. Tang, R.G. Yang, H. Lee, D.Z. Wang, Z.F. Ren, J.P. Fleurial, and P. Gogna, in press.

[8] I. Terasaki, Y. Sasago, K. Uchinokura, *Phys. Rev.* B **56** 12685-12687 (1997).

[9] M. Mikami, R. Funashashi, M. Yoshimura, Y Mori, and T. Sasaki, *J. Appl. Phys.* **94** (10) 6579 - 6582 (2003).

[10] M. Mikami, R. Funahashi, *J. Solid State Chem.* **178** 1670-1674 (2005).

[11] D. Grebille, S. Lambert, F. Bouree, and V. Petricek, *J. Appl. Crystallogr.* **37** 823-831 (2004).

[12] A.C. Masset, C. Michel, A. Maignan, M. Hervieu, O. Toulemonde, F. Studer, and B. Raveau, *Phys. Rev* B **62** 166-175 (2000).

[13] H. Minami, K. Itaka, H. Kawaji, Q.J. Wang, H. Koinuma, and M. Lippmaa, *Appl. Surface Sci.* **197** 442-447 (2002).

[14] W. Wong-Ng, Y.F. Hu, M.D. Vaudin, B. He, M. Otani, N.D. Lowhorn, and Q. Li, J. Appl. Phys, **102**(3) 33520 (2007).

[15] Bruker User Manual, General Area Detector Diffraction System (GADDS), vs. 4.9, 1999, Bruker AXS, Inc. Madison, WI 53711.

[16] M. Otani, N.D. Lowhorn, P.K. Schenck, W. Wong-Ng, and M. Green: Appl. Phys. Lett., **91** (2007) 132102.

[17] M. Otani, K. Itaka, W. Wong-Ng, P.K. Schenck, and H. Koinuma, Appl. Surface Science, **254** 765-767 (2007).

[18] Physical Property Measurement system (PPMS), manufactured by Quantum Design, San Diego, CA. 92121-3733, USA.

[19] M.D. Vaudin, M.W. Rupich, M. Jowett, G.N. Riley, Jr., J.F. Bingert, J Mater Res. **13** 2910 (1998).

[20] Y.F. Hu, W.D. Si, E. Sutter, and Q. Li, *Appl Phys. Lett.* **86** 082103 (2005).

[21] T. Takami, H. Ikuta, and U. Mizutani, Jap. J. Appl. Phys. 43 (22) 8208-8212 (2004).

[22] H. Fjellvag, E. Gulbrandsen, S. Aasland, A. Olsem, B. C. Hauback, J. Solid State Chem. **124** 190 (1996).

[23] M. Mikami and R. Funahashi, IEEE Proceedings on 22nd International Conference on Thermoelectrics, p. 200, (2003)

[24] R.D. Shannon, Acta Crystallogr. **A32**, 751-767 (1976).

PREPARATION OF LOW-LOSS TITANIUM DIOXIDE FOR MICROWAVE FREQUENCY APPLICATIONS

L. Zhang, K. Shqau, and H. Verweij
Group Inorganic Materials Science, Department of Materials Science and Engineering, The Ohio State University, Columbus OH, USA

G. Mumcu, K. Sertel, J.L. Volakis
ElectroScience Laboratory, Department of Electrical and Computer Engineering, The Ohio State University, Columbus OH, USA

ABSTRACT

Dense titanium dioxide (TiO_2) was prepared from commercially available high-purity powders. The powders were first deagglomerated and colloidally stabilized in aqueous NH_3 to form a stable suspension. The suspension was screened, followed by colloidal-filtration, drying and sintering. Sintering at 1000°C for 10 hours resulted in > 99.5% density and 2.2 μm grains. The dielectric loss of the dense-sintered TiO_2 was 1.4×10^{-4} at 6.4 GHz at room temperature. This low dielectric loss is attributed to a low sintering temperature and homogeneous microstructure. Dense, low-loss TiO_2 is particularly interesting for microwave applications because of its very high dielectric constant at GHz frequencies. The process presented here will be used to make dense TiO_2 parts with complex shape in new microwave antenna concepts, for its high dielectric constant and low dielectric loss at microwave frequencies.

INTRODUCTION

Titanium dioxide (TiO_2, rutile phase) is considered for use in new microwave devices, because of its high permittivity (dielectric constant $\varepsilon_r = 115$, measured in this study), and low dielectric loss ($\tan\delta$) at microwave frequencies [1]. One drawback of the application of TiO_2 is a strong temperature-dependence of properties which has to be addressed by temperature control of the eventual device [2]. Dense TiO_2 is recently proposed as a constituent of a recent innovative photonic assembly (PA) antenna design. A prototype for this design, built with α-$Al_2O_3|BaTiO_3$ laminates, was shown to exhibit the expected directive radiation properties at GHz frequencies [3]. To further improve the antenna gain, $BaTiO_3$ has to be replaced by dense TiO_2, which has a higher ε_r and lower $\tan\delta$. In this study we made an effort to prepare such dense and low-loss TiO_2 using controlled colloidal processing.

Traditional powder compaction resulted in TiO_2 densification temperatures of 1200…1500°C with several hours of soak time [1,4,5]. Such high sintering temperatures lead to oxygen deficiency, resulting in high $\tan\delta$ due to resistive dissipation by mobile electronic charge carriers [2,6]. In addition, poor mechanical strength is often observed in such samples due to substantial grain growth (grain size ~20 μm) and inhomogeneous microstructures. Dopants have been applied to reduce $\tan\delta$ by compensating the reduction effect [1,6-8], but with little improvement on microstructures or decrease of sintering temperatures. Low densification temperature and fine-grain microstructures can be achieved by processing monodisperse particles from controlled hydrolysis of titanium alkoxides [9,10]. Although no microwave dielectric measurements have been reported for such TiO_2 samples, their low sintering temperature and homogeneous microstructure of small grains (1-2 μm) are expected to lead to a low $\tan\delta$ [6,11]. However, the high cost of alkoxides and required high purity likely inhibit large scale production of TiO_2 by this route.

This paper reports on low-temperature sintering of polycrystalline TiO_2 (rutile) with low dielectric loss, directly from inexpensive commercial powders of very high purity. Here chemical purity is considered prior to phase purity in the starting powders, because chemical impurities may persistently affect dielectric loss and/or microstructure homogeneity in the sintered TiO_2 samples. In the present study we show how the problems of several TiO_2 phases and less-than-optimum particle morphology can be addressed to obtain good results through a viable process.

EXPERIMENT

Sample preparation

TiO_2 powders (Ishihara Corp.) were used as received, with >99.99% purity, 72.7 wt% rutile phase and an average particle size of 0.15 μm. Suspensions with 14 vol% TiO_2 solid load were prepared using ultrasonic dispersion, with a digital sonifier (Branson Ultrasonics Corp.) in aqueous NH_3 at $p_H = 10.5$. For a batch of 100 g TiO_2 suspension, the ultrasonic treatment was carried out with 70 watts power for 8 min in a 100 ml double-wall beaker. The beaker was water-cooled to avoid excessive heating by ultrasonic energy dissipation [12]. The TiO_2 suspension was screened to remove any big agglomerates and foreign contaminations using a Nylon Spectra Mesh® with (a) 5 μm aperture, 2% opening area and 100 μm thickness or (b) 10 μm aperture, 2% opening area and 45 μm thickness (Spectrum Laboratories, Inc.). Slight ultrasonic agitation was applied at 4 watts power to avoid clogging of meshes. Green compacts (disks) were formed by colloidal filtration of the suspension onto a polyethersulfone membrane with 0.22 μm pore size (Millipore Corp.). After drying overnight, TiO_2 green compacts were calcined for 10 hours at 600°C to 1100°C, with a 5°C/min heating and cooling rate.

Sample characterization

Zeta(ζ)-potential measurements were performed on 1.25 vol% TiO_2 aqueous suspensions using a Zetaprobe Analyzer™ (Colloidal Dynamics Inc.). The titrants were (a) electrostatic stabilizers: 5 mol/L HNO_3 and 5 mol/L tetramethyl ammonium hydroxide (TMAH), and (b) electrosteric stabilizer: 1 mol/L Aluminon solutions [12]. To further study its stabilization in aqueous HNO_3 and NH_3 solutions, 12.5 vol% TiO_2 suspensions were prepared in aqueous HNO_3 with $p_H = 2.0$, 2.4 and 2.7, and NH_3 with $p_H = 9.4$, 10.5 and 11.1. The p_H of the solutions, TiO_2 suspensions and gels was measured using a p_H/ISE meter (Model 710A, Orion).

X-ray diffraction (XRD) was performed on both as-received powders and thermal-treated TiO_2 samples, using a Scintag XDS2000 diffractometer with Cu-Kα radiation ($\lambda = 1.5406$ Å) with $2\theta = 20...80°$. Scanning electron microscopy (SEM) was carried out using a Field-Emission Environmental SEM Philips XL30 (Eindhoven, the Netherlands) on thermally etched cross-sections of TiO_2 samples sintered at 900...1100°C. The density of calcined samples was measured using a mercury pyconometer (Model DAB100-1, Porous Materials Inc.) at room temperature. 300°C calcined samples, with identical XRD patterns as the as-received powders, were used to determine the green density. For each treatment temperature, three samples were measured to obtain an average density, and a 95% confidence interval. The average grain size of TiO_2 sintered at 900...1100°C was characterized using a linear intercept technique applied to nontextured grains of tetrakaidecahedral shape [13]. At least 200 grain intercepts were measured for each sample to obtain a sufficient accuracy.

For microwave loss characterization, sintered TiO_2 samples were machined to $3.75 \times 12.50 \times 2.00$ mm^3 rectangular dimensions by Louwers Glass and Ceramic Technologies (the

Netherlands). The complex dielectric constant ($\varepsilon_r = \varepsilon' + j\varepsilon''$) of these samples was measured using a resonant electromagnetic cavity coupled with finite element (FE) simulation [14]. The copper cavity was fabricated with 24×12.8×80 mm^3 rectangular dimensions, proportional to those of the samples. It was designed to support resonant electromagnetic field modes with the strong electromagnetic field concentrated within the sample at the center of the cavity. The cavity was connected to a vector network analyzer (Agilent E8362B) and the return loss (S_{11}) was recorded. The quality factor of the sample-loaded cavity was used to determine dielectric loss. After conductor loss of the copper cavity was mathematically factored out, dielectric loss ($\tan \delta = \varepsilon''/\varepsilon'$) in TiO_2 samples was characterized with an accuracy of $\Delta(\tan \delta) = 10^{-7}$. A numerical search was carried out to match the measured resonance frequencies with those predicted from FE simulations to determine ε'.

RESULTS AND DISCUSSIONS

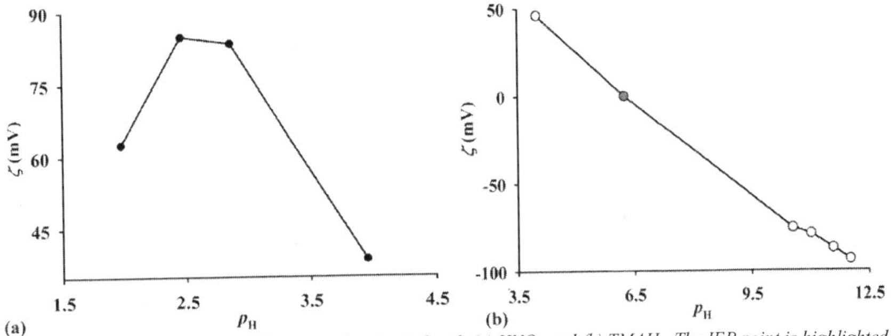

Figure 1: ζ potential of 1.25 vol% TiO_2 suspension titrated with (a) HNO_3 and (b) TMAH. The IEP point is highlighted in red.

The isoelectric point (IEP) of the TiO_2 powders was measured to be at $p_H = 6.2$. The ζ potential of 1.25 vol% TiO_2 suspension reached a maximum of +85 mV at $p_H = 2.5$ and -93 mV at $p_H = 12.0$, as shown in figure 1. On the contrary, $|\zeta|$ was always < 60 mV in the presence of Aluminon stabilizer, even at a concentration up to 0.13 wt%. Though Aluminon may show a high $|\zeta|$ in basic environment [12], stabilization by solely H^+ or OH^- was preferred to avoid another, possibly contaminating additive. The decrease of ζ potential at $p_H < 2.3$ may be ascribed to dissolution of TiO_2 in HNO_3; this was not observed at high p_H values. To establish a practical p_H range, TiO_2 suspensions with a solid load of 12.5 vol%, was prepared with HNO_3 solution with $p_H = 2.0...2.7$, and NH_3 solution with $p_H = 9.4...11.1$ respectively. Figure 2 shows p_H in the starting HNO_3, NH_3 solutions, and the resultant TiO_2 suspensions. It was found that gelation occurred when the initial solutions had $2.6 < p_H < 9.5$. Sedimentation experiments were carried out for the remaining p_H ranges. The suspensions, prepared from $p_H = 2.0$ HNO_3 and $p_H = 10.5$ NH_3 solution were observed to be the most stable at $p_H = 2.3$ and 10.0 respectively. Since the acidic $p_H = 2.3$ is close to the p_H where TiO_2 starts to dissolve, a $p_H = 10.0$ was used to prepare green compacts. Both 5 and 10 μm Spectra Mesh® screens were tested for purification of TiO_2 suspensions, which resulted in similar microstructures in the final calcined samples. The 5 and 10 μm meshes had an effective aperture of 3 and 9 μm respectively as observed by SEM. 10 μm meshes were used routinely because screening with 5 μm meshes required an unacceptably long treatment time.

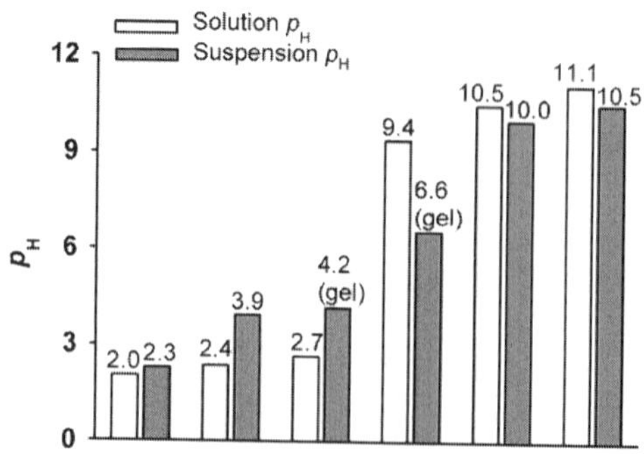

Figure 2: p_H in the starting HNO_3, NH_3 solutions, and the resultant 12.5 vol% TiO_2 suspensions.

The solid load in the TiO_2 suspension was optimized to achieve both high green density (ρ_g) and efficient colloidal-filtration. Using $p_H = 10.5$ NH_3 solution as dispersion medium, TiO_2 suspensions with >20 vol% solid load showed a p_H decreasing to 9.2, and sedimentation occurred within minutes. A lower solid load generally led to a higher ρ_g, but this, in turn, required more suspension volume and longer filtration time for dense samples with certain mass and dimensions. For example, the required filtration times for 25 ml 10.0 vol%, 20 ml 12.5 vol% and 13 ml 19 vol% TiO_2 suspensions were >6 hrs, 4.5-5 hrs and 2.5 hrs, respectively. Finally 17 ml 14 vol% TiO_2 suspension with 3.5 hrs' filtration was chosen as a compromise between high ρ_g and filtration efficiency (targeting dense TiO_2 of $3.75\times12.50\times2.00$ mm^3 for in-cavity measurement). After filtration and drying, the green compacts remained intact with sufficient strength for handling and a smooth shiny surface, as shown in figure 3.

Figure 3: Side view of a TiO_2 green body (diameter of 42.3 mm, thickness of 2.3 mm) after drying overnight.

The as-received TiO_2 powders consisted of mainly rutile phase (72.7 wt% from manufacturer data) and anatase phase, as shown in the XRD patterns in figure 4. The average density was estimated to be 4.16 g/cm^3 accordingly. The relative green density of compacts obtained at optimum conditions was therefore calculated to be 57.5±0.7%. After treatment at >700°C, the anatase phase completely transformed to rutile, as confirmed by XRD analysis. The dominant as-received impurities were 25 ppm Fe_2O_3, 10 ppm Nb_2O_5 and 10 ppm Na_2O. This impurity concentration was at least 10 times lower than the commonly used 500 ppm for single dopants. Therefore the commercial TiO_2 powders were considered suitable for a study of the dielectric properties of unintentionally doped polycrystalline rutile TiO_2.

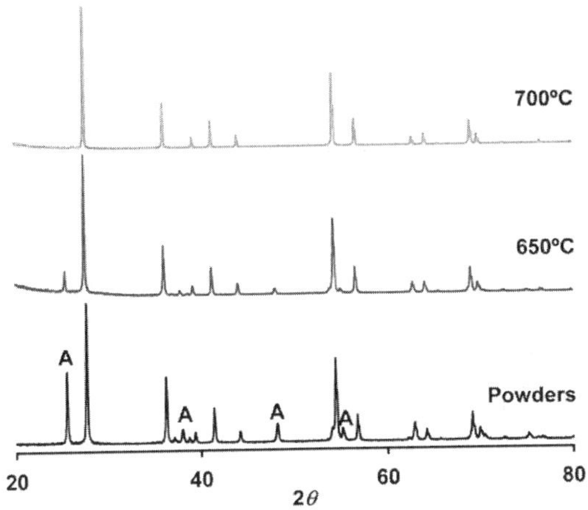

Figure 4: XRD patterns of the as-received TiO_2 powders and 650°C, 700°C calcined samples. The diffraction peaks from anatase phase are indicated by "A".

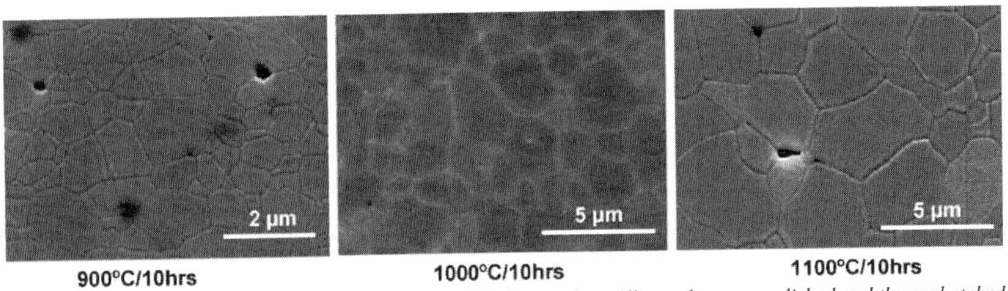

Figure 5: SEM images of 900...1100°C calcined TiO_2 samples. All samples were polished and thermal etched.

As shown in figure 5, TiO_2 grains grew rapidly from 1.1 μm, 2.2 μm to 4 μm when the sintering temperature increased from 900 to 1100°C. The faceted pores in TiO_2, sintered at 900°C were attributed to break out of grains during polishing of the relatively weak, porous samples. As the sintering temperature increased, the relative density (ρ_r) of TiO_2 increased and reached >99.5% at 1000°C, as shown in figure 6. This densification temperature is, to our knowledge, the lowest for commercial undoped TiO_2 powders. It was achieved primarily because of the optimized homogeneity of the TiO_2 green compact. A higher temperature, however, resulted in a small but significant decrease in ρ_r. This is attributed to desintering by anomalous grain growth and/or expansion by oxygen released from TiO_2 reduction [15]. From in-cavity measurement at 6.3...6.4 GHz, TiO_2 sintered at 900...1100°C showed a increasing dielectric constant from 110 to 115, and a decreasing tan δ, from 1.8×10^{-4} to 1.4×10^{-4}, also shown in figure 6. The tan δ of 1.4×10^{-4} is 5 times lower than the value at 3 GHz reported for undoped TiO_2 prepared from sintering pressed pellets [1,16]. The measurement frequency of 6.4 GHz was higher than that used to obtain the values reported earlier [1,4] and dielectric losses tend to increase with microwave frequencies. Dielectric properties of TiO_2 samples with ρ_r <96% were not measured because their relatively large porosity resulted in a lower permittivity, and a high tan δ due to energy dissipation by polar species absorbed on the pore surface. The dielectric loss in this study can be ascribed to random-orientation of larger grains, defects in the surface of grains and residual pores, and contamination of the exterior sample surfaces.

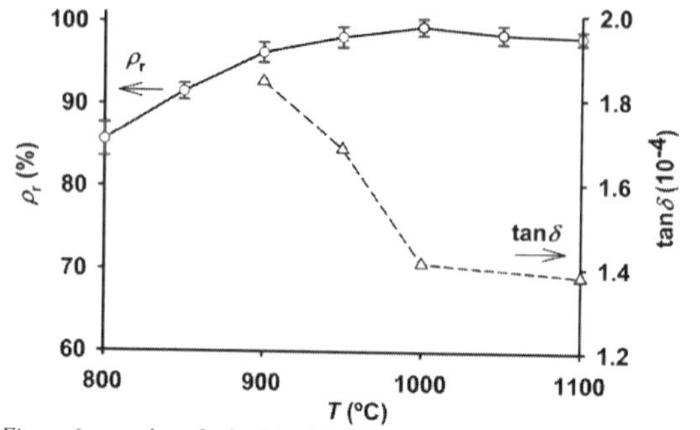

Figure 6: ρ_r and tan δ of calcined TiO_2 versus the sintering temperature.

Conclusions

Commercially available TiO_2 powders were colloidally stabilized and pressure-filtered to form green compacts with a high relative density of 57.5%. Such compacts densified almost completely at a temperature as low as 1000°C. Thus obtained sintered TiO_2 had a homogeneous micro-structure, a small grain size of 2.2 μm, and a low dielectric loss of 1.4×10^{-4} at 6.4 GHz. Both the sintering temperature and room temperature dielectric loss are the lowest among all the reported values for commercial undoped TiO_2 samples. The obtained low-loss TiO_2 will be used to build microwave antennas because of its very high dielectric constant at GHz ferquencies. We expect that the method as presented can be a good starting point to develop viable industrial manufacturing protocols, such as inkjet printing of TiO_2 suspension, and tape-casting of TiO_2 slurries.

ACKNOWLEDGMENTS

This work was supported by the U.S. Air Force Office of Scientific Research under the grant FA9550-04-1-0359, and the Ohio State University Office of Research Large Interdisciplinary Grant Program.

REFERENCES

[1] A. Templeton, X. Wang, S.J. Penn, S.J. Webb, L.F. Cohen and N.M. Alford, Microwave dielectric loss of titanium oxide," *J. Am. Ceram. Soc.*, Vol 83 (1), 2000, p 95-100.

[2] R.A. Parker, Static dielectric constant of rutile (TiO_2), 1.6-1060°K, *Phys. Rev.*, Vol 124, 1961, p 1719-1722.

[3] L. Zhang, G. Mumcu, S. Yarga, K. Sertel, J.L. Volakis and H. Verweij, Fabrication and characterization of anisotropic dielectrics for low-loss microwave applications, *J. Mater. Sci.*, Vol 43 (5), 2008, p 1505-1509.

[4] L. Egerton and J. Thomson Jr., Preparation of high density ceramic TiO2 having low dielectric loss at microwave frequencies, *Ceram. Bul.*, Vol 50 (11), 1971, p 924-928.

[5] T. Burg, T. Bak, J. Nowotny, L. Sheppard, C.C. Sorrell and E.R. Vance, Effect of sintering on microstructure of TiO2 ceramics, *Adv. Appl. Ceram.*, Vol 106 (1-2), 2007, p 57-62.

[6] F.A. Grant, Properties of rutile (titanium dioxide), *Rev. Mod. Phys.*, Vol 31, 1959, p 646-674.

[7] J.F. Baumard and E. Tani, Electric conductivity and charge compensation in Nb-doped TiO_2 rutile, *J. Chem. Phys.*, Vol 67 (3), 1977, p 857-860.

[8] L.G. Rowan, Microwave dielectric loss of V^{+4}-doped TiO_2 at low temperatures," *Phys. Rev. B.*, Vol 5 (5), 1972, p 1675-1676.

[9] M.F. Yan and W.W. Rhodes, Low temperature sintering of TiO_2, *Mater. Sci. Eng.*, Vol 61, 1983, p 59-66.

[10] E.A. Barringer and H.K. Bowen, Formation, packing, and sintering of monodisperse tita-nium dioxide powders, *J. Am. Ceram. Soc.*, Vol 65 (12), 1982, p C199-201.

[11] S.J. Penn, N.M. Alford, A. Templeton, X. Wang, M. Xu, M. Reece and K. Schrapel, Effect of porosity and grain size on the microwave dielectric properties of sintered alumina, *J. Am. Ce-ram. Soc.*, Vol 80 (7), 1997, p 1885-1888.

[12] K. Shqau, M.L. Mottern, D. Yu, and H. Verweij, Preparation and properties of porous α-Al_2O_3 membrane supports, *J. Am. Ceram. Soc.*, Vol 89 (6), 2006, p 1790–1794.

[13] M.I. Menderson, Average grain size in polycrystalline ceramics, *J. Am. Ceram. Soc.*, Vol 52 (8), 1969, p 443-446.

[14] G. Mumcu, K. Sertel, and J.L. Volakis, A measurement process to characterize natural and engineered low loss uniaxial dielectric materials at microwave frequencies, *IEEE T. on Microw. Theory.*, Vol 56 (1), 2008, p 217-223.

[15] O. Sudre and F.F. Lange, The effect of inclusions on densification: III, the desintering phe-nomenon, *J. Am. Ceram. Soc.*, Vol 75 (12), 1992, p 3241-3251.

[16] C.T. Dervos, E. Thirios, J. Novacovich, P. Vassiliou, and P. Skafidas, Permittivity proper-ties of thermally treated TiO_2, *Mater. Lett.*, Vol 58, 2004, p 1502-1507.

ANALYTIC METHODS FOR DETERMINATION OF ACTIVATION ENERGY USING THE MASTER SINTERING CURVE APPROACH

Matthew Schurwanz and Stephen J. Lombardo*
Department of Chemical Engineering
University of Missouri
Columbia, MO 65211, USA

*Department of Mechanical and Aerospace Engineering
University of Missouri
Columbia, MO 65211, USA

ABSTRACT

An approximation developed by Lee and Beck is applied to evaluate the integral that appears in the Master Sintering Curve (MSC) approach. With this approximation, two analytical equations are derived that can be used to determine the activation energy of sintering using only temperature and density data. The two analytical expressions are compared with numeric integration on simulated density versus temperature data for three values of the activation energy of 250, 450, and 650 kJ/mol. The two analytical equations lead to values for activation energy within 0.08% of the value used to create the data.

INTRODUCTION

We have recently developed a combined dilatometer-mass spectrometer (CDMS) apparatus for simultaneously monitoring the sintering process and the accompanying high temperature gas-phase chemistry. With this apparatus, experiments are conducted at different linear heating rates and both the kinetics of sample shrinkage and appearance of gas-phase species are recorded. Both sets of the data can then be analyzed in order to obtain the activation energy for the different kinetic processes.

For the case of the sintering of ceramic materials, the calculation of activation energies often requires either complex experimental techniques or the assumption of simplified models. One integral method of kinetic analysis, the Master Sintering Curve (MSC) approach[1], is advantageous for calculating the activation energy of sintering because no mechanistic models is assumed and the only required data are density versus temperature curves obtained at different linear heating rates.[2-6] In the standard MSC approach, however, an integral appears which must be evaluated numerically.

In this work, we derive two analytical equations for evaluating the integral which appears in the MSC approach. We first use an approximation developed by Lee and Beck,[7] and then modify this for the specific form of the integral which appears in the MSC formulation. These two methods are then compared with a numerical integration based on the trapezoidal rule.

MODEL

Method of Lee and Beck

The analytical evaluation of the integral appearing in the MSC approach is based on the methodology of Lee and Beck[7], who proposed an analytical integral approximation for kinetic weight loss data obtained by thermogravimetric analysis (TGA). Although originally applied to coal conversion, their technique is also applicable for a wide variety of kinetic processes described by an activated (Arrhenius-type) exponential dependence on temperature.

For a solid, the activated kinetics of decomposition are typically represented by:

$$\frac{d\omega}{dt} = A\exp\left(\frac{-Q}{RT}\right)f(\omega) \tag{1}$$

where ω is the decomposed fraction of solid at time t, $f(\omega)$ is a function which depends on the reaction mechanism, A is the preexponential frequency factor, Q is the activation energy, R is the universal gas constant and T is the absolute temperature. For a linear heating rate, β, the relationship between time and temperature is given by:

$$\beta \equiv \frac{dT}{dt} \tag{2}$$

Equations (1) and (2) can be combined, rearranged, and then integrated between the initial and final temperatures, T_0 and T, and initial and final conversion, ω_0 and ω, so that

$$\int_{\omega_0}^{\omega}\frac{d\omega}{f(\omega)} = \frac{A}{\beta}\int_{T_0}^{T}\exp\left(-\frac{Q}{RT}\right)dT = \frac{ART^2}{\beta Q}\exp\left(-\frac{Q}{RT}\right)\Bigg|_{T_0}^{T} - \int_{T_0}^{T}\frac{2RT}{Q}\exp\left(-\frac{Q}{RT}\right)dT \tag{3}$$

The expression on the right-hand side of Equation (3) is the result of integration by parts, which again leads to the same exponential function that cannot be evaluated analytically. To circumvent this recursive behavior, Equation (3) can be rearranged so that a value of $(1+2RT/Q)$ appears in the integral. For moderate temperatures and large activations energies, $2RT/Q$ is far less than unity, thus making $(1+2RT/Q)$ approximately equal to one and therefore approximately constant. When $(1+2RT/Q)$ is factored from the integral, Equation (3) can be rearranged further to give:

$$\int_{T_0}^{T}\exp\left(-\frac{Q}{RT}\right)dT = \frac{\left[\frac{RT^2}{Q}\exp\left(-\frac{Q}{RT}\right)\right]}{1+\frac{2RT}{Q}}\Bigg|_{T_0}^{T} \tag{4}$$

which leads to:

$$\int_{\omega_0}^{\omega} \frac{d\omega}{f(\omega)} = \frac{A}{\beta}\left[\frac{\left(\frac{RT^2}{Q}\right)\exp\left(-\frac{Q}{RT}\right)}{\left(1+\frac{2RT}{Q}\right)} - \frac{\left(\frac{RT_0^2}{Q}\right)\exp\left(-\frac{Q}{RT_0}\right)}{\left(1+\frac{2RT_0}{Q}\right)}\right] \equiv \frac{A}{\beta}\left(\Gamma_T - \Gamma_{T_0}\right) \tag{5}$$

where Γ_T and Γ_{T_0} are lumped expressions to denote the final Lee and Beck equation. Γ_{T_0} is dependent on the initial temperature, and is a constant if Q is known.

The accuracy of the Lee and Beck method depends on the approximation that $(1+2RT/Q)$ is constant, which is most accurate for high activation energies, low temperatures, and for small ranges in temperature. For sintering, these conditions are generally satisfied because even though the temperatures are high, approximately 1500 – 1700 Kelvin, the temperature range is relatively narrow, and the activation energy can be as large as 500-700 kJ/mol.

To use Equation (5) to obtain the activation energy of sintering, ω is replaced by ρ, the fractional density of the sample relative to the maximum theoretical density. Equation (5) can then be compactly represented using the following definition for $F(\rho)$:

$$F(\rho) \equiv \int_{\rho_0}^{\rho} \frac{d\rho}{f(\rho)} = \frac{A}{\beta}\int_{T_0}^{T} \exp\left(-\frac{Q}{RT}\right)dT = \frac{A}{\beta}\left(\Gamma_T - \Gamma_{T_0}\right) \tag{6}$$

Different models for $f(\rho)$ exist, and some common models and their integrated forms are shown in Table 1.

Table 1. Kinetic mechanisms with their integrated forms.

Kinetic Model	$f(\rho)$	$F(\rho)$
Zero order	$(1-\alpha)^0$	$\alpha - \alpha_0$
First order	$(1-\alpha)^1$	$\ln(1-\alpha_0) - \ln(1-\alpha)$
Second order	$(1-\alpha)^2$	$1/(1-\alpha) - 1/(1-\alpha_0)$

Equation (6) can be further simplified using the approximation that the value of Γ_{T_0} is small when compared to Γ_T which leads to

$$F(\rho) = \frac{A}{\beta}\Gamma_T = \frac{A}{\beta}\frac{\left(\frac{RT^2}{Q}\right)\exp\left(-\frac{Q}{RT}\right)}{\left(1+\frac{2RT}{Q}\right)} \tag{7}$$

Equation (7) can next be rearranged to

$$\ln\left[\frac{F(\rho)}{T^2}\right] = \ln\left[\frac{AR}{\beta(Q+2RT)}\right] - \frac{Q}{RT} \tag{8}$$

When $\ln\left[F(\rho)/T^2\right]$ is plotted vs. $1/T$, the slope is equal to $-Q/R$ and the intercept is equal to $\ln\left[AR/\beta(Q+2RT)\right]$. Although the intercept can be used to determine the pre-exponential factor graphically, it is more commonly found by averaging values of A obtained algebraically from an equation such as Equation (6) or (8) along with the known value of Q obtained from the slope. Equation (8) and the above procedure represents use of the method of Lee and Beck to obtain activation energies.

Master Sintering Curve Method

Hansen *et al.*[8] developed a combined stage sintering model that could be used to describe the sintering process as:

$$-\frac{dL}{Ldt} = \frac{\gamma\,\Omega}{k_b T}\left(\frac{\Gamma_v D_v}{G^3} + \frac{\Gamma_b \delta\, D_b}{G^4}\right) \tag{9}$$

where γ is the surface energy, Ω is the atomic volume, k_b is the Boltzmann constant, T is the absolute temperature, G is the mean grain diameter, D_v and D_b are the coefficients for volume and grain-boundary diffusion, respectively, and δ is the width of grain boundary. The quantities Γ_v and Γ_b are lumped scaling parameters, later referred to as Γ_n, which relate driving forces, mean diffusion distances, and other geometric features of the microstructure which influence the sintering rate in terms of the mean grain diameter. These values are more thoroughly explained in the paper of Hansen *et al.*[8] When there is only moderate diffusion and grain growth, the values of Γ_n are dependent on the density but independent of the heating schedule. Although it is possible to obtain all of the values from experiments or theoretically with simplified models, the calculated values and those found experimentally are usually significantly different. In addition, the process of experimental determination is quite complex and not practical to do on an industrial scale.

To circumvent these difficulties, Su and Johnson[1] used the combined stage sintering model to develop the master sintering curve (MSC) methodology to obtain the activation energy of sintering. This method requires only density and temperature data from sintering experiments conducted at multiple, different linear heating rates.

When diffusion is dominated by either grain boundary or volume diffusion, Γ_n and G are only functions of density. Based on this assumption, Equation (9) can be rearranged so that the left-hand side (lhs) has all properties dependant on the microstructure and the right-hand side (rhs) is dependant only on the heating profile and activation energy, as given by:

$$lhs = \Phi(\rho) \equiv \frac{k_b}{\gamma\,\Omega D_0}\int_{\rho_0}^{\rho}\frac{(G(\rho))^n}{3\rho\Gamma(\rho)}\,d\rho \tag{10}$$

$$rhs = \Theta(t, T(t)) \equiv \int_0^t \frac{1}{T} \exp\left(-\frac{Q}{RT}\right) dt \tag{11}$$

$$\Phi(\rho) = \Theta(t, T(t)) \tag{12}$$

Because of the underlying assumptions, Equations (10) and (11) are only valid when the microstructural evolution is dependent on density alone, which is most true for the relative density range of 0.6 to 0.9. The relationship between ρ and $\Phi(\rho)$ is the definition of the MSC, which is unique for each powder and green body process, including fixed green density. Since Equation (11) is a function of only time and temperature, both of which can be monitored experimentally, $\Theta(t, T(t))$ is known if Q is known or can be approximated. We can then use the equality in Equation (12) to obtain $\Phi(\rho)$ without determining the individual factors comprising Equation 10 or using any further simplifications. Because Equation (11) is dependant on time and heating rate, but Equation (10) is not, when $\Theta(t, T(t))$ is plotted vs. ρ using experimental data collected at different heating rates, the graphs will collapse to a single curve only when the true activation energy is used. In practice, the natural log of Equation (11) is often used.

To find the activation energy of sintering with the MSC method, it is not necessary for the mechanism underlying $\Phi(\rho)$ be known. All that is required is that the mechanism not change over the sintering range studied and be independent of changes in temperature. With this in mind, $\Phi(\rho)$ can be expressed by the general equation for $F(\rho)$ as defined in Equation (7). Equations (1), (2), (11) and (12) can now be expressed as:

$$F(\rho) = \int_{T_0}^{T} \frac{1}{\beta T} \exp\left(-\frac{Q}{RT}\right) dT \tag{13}$$

This expression emphasizes that no knowledge of F is required, except the dependence on density. Equation (13) is fundamentally similar to the Arrhenius dependence in Equation (3). This suggests that, with slight alterations, the general MSC method could be applied to find the activation energies of a wide variety of thermally activated processes.

Application of the Lee and Beck Approximation to the MSC Method

For sintering, the activation energy is often between 200-700 kJ/mol while the majority of density change occurs over a range of approximately 400 degrees C. These two factors make the Lee-Beck approximation suitable for use with the Master Sintering Curve. However, Equation (11) must be handled differently from Equation (1) due to the presence of $1/T$ in the integral of Equation (11).

First, the integral in Equation (11) must be set in terms of T using the linear heating rate, β, defined in Equation (2). Then integration by parts gives:

$$\Theta(T(t)) = \frac{1}{\beta} \int_{T_0}^{T} \frac{1}{T} \exp\left(-\frac{Q}{RT}\right) dT = \frac{1}{\beta}\left(\frac{1}{T} \int_{T_0}^{T} \exp\left(-\frac{Q}{RT}\right) dT - \int_{T_0}^{T}\left[\int_{T_0}^{T} \exp\left(-\frac{Q}{RT}\right) dT\right]\left(-\frac{1}{T^2}\right) dT\right) \tag{14}$$

where the integral from the Lee and Beck equation now appears twice. Combining Equation (14) with the definitions in Equation (5) and simplifying yields:

$$\Theta(T(t)) = \frac{1}{\beta}\left(\frac{\Gamma_T}{T} + \frac{R}{(Q+2RT)}\int_{T_0}^{T}\exp\left(-\frac{Q}{RT}\right)dT\right) \tag{15}$$

Once again, this contains the integral from the Lee and Beck expression. Additional use of Equation (5) yields the final analytical MSC equation:

$$\Theta(T(t)) = \frac{1}{\beta}\left(\frac{\Gamma_T}{T} + \left(\Gamma_T - \Gamma_{T_0}\right)\frac{R}{Q}\frac{1}{\left(1+2RT/Q\right)}\right) \tag{16}$$

which can also be expressed as:

$$\Theta(T(t)) = \frac{1}{\beta}\left(\frac{\left(\frac{RT}{Q}\right)\exp\left(-\frac{Q}{RT}\right)}{\left(1+\frac{2RT}{Q}\right)} + \left[\frac{\left(\frac{RT^2}{Q}\right)\exp\left(-\frac{Q}{RT}\right)}{\left(1+\frac{2RT}{Q}\right)} - \frac{\left(\frac{RT_0^2}{Q}\right)\exp\left(-\frac{Q}{RT_0}\right)}{\left(1+\frac{2RT_0}{Q}\right)}\right]\frac{R}{\left(Q+2RT\right)}\right) \tag{17}$$

All of the previous equations for the master sintering curve can now be summarized as:

$$\frac{k_b}{\gamma\Omega D_0}\int_{\rho_0}^{\rho}\frac{(G(\rho))^n}{3\rho\Gamma(\rho)}d\rho \equiv \Phi(\rho) = \Theta(t,T(t)) \equiv \int_0^t\frac{1}{T}\exp\left(-\frac{Q}{RT}\right)dt \approx \frac{1}{\beta}\left(\frac{\Gamma_T}{T} + \left(\Gamma_T - \Gamma_{T_0}\right)\frac{R}{Q}\frac{1}{\left(1+2RT/Q\right)}\right) \tag{18}$$

The last term is only approximately equal to the others due to three uses of the Lee and Beck approximation.

RESULTS AND DISCUSSION

To use Equation 18, density versus temperature data were simulated for the sintering process at heating rates of 20, 12, and 5°C/min. These data were generated using second order kinetics from

$$\frac{1}{1-\rho} - \frac{1}{1-\rho_0} = \frac{A}{\beta}\left(\frac{\Gamma_T}{T} + (\Gamma_T - \Gamma_{T_0})\frac{R}{E}\frac{1}{1+2RT/Q}\right) \tag{19}$$

Equation (19) is thus a slightly modified form of the analytical MSC expression, Equation (16), where a pre-exponential factor, A, was added in an attempt to simulate the lack of clear knowledge of the values represented in $\Phi(\rho)$ from Equation (10). This also insures that the simulation will not be analyzed with the same equation used to generate it. For all simulations, A was equal to $8\text{x}10^{14}$ /min with Q set equal to 250, 450, and 650 kJ/mol. An example of the simulated density data versus temperature data for Q=450 kJ/mol is shown in Fig. 1, in which the form of the sintering data is qualitatively similar to what is obtained from real ceramic samples.

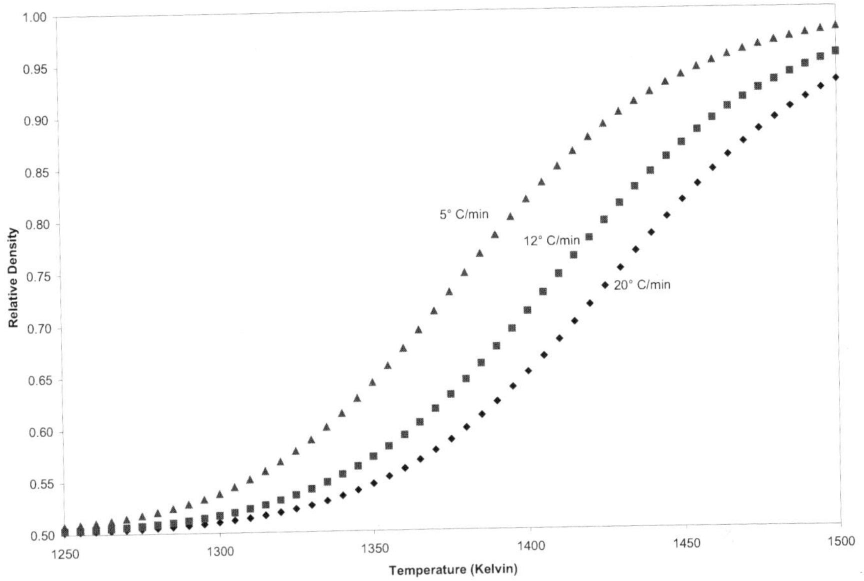

Figure 1: Data simulated using second order kinetics from Equation (19). Three different liners heating rates were evaluated for an activation energy of 450 kJ/mol.

For each set of simulated data as in Fig. 1, the activation energy was found by three different methods. In the first, Equation (13) was integrated numerically using the trapezoidal rule with steps of 0.001 relative density. For the second, the analytical equation given by Equation (16) was used. Finally, another approximated analytical equation was used with the assumption that the value of $1/T$ in Equation (13) is relatively constant when compared to the rate of change in the exponential and so can be averaged and removed from integral. For convenience, the three approaches used here are repeated as Equations (20), (21), and (22):

$$\ln[\Phi(\rho)] = \ln\left[\int_{T_0}^{T} \frac{1}{\beta T} \exp\left(-\frac{Q}{RT}\right) dT\right] \tag{20}$$

$$\ln[\Phi(\rho)] = \ln\left[\frac{1}{\beta}\left(\frac{\Gamma_T}{T} + \left(\Gamma_T \quad \Gamma_{T_0}\right)\frac{R}{Q}\frac{1}{(1+2RT/Q)}\right)\right] \tag{21}$$

$$\ln[\Phi(\rho)] = \ln\left[\frac{1}{\beta \overline{T}}\left(\Gamma_T - \Gamma_{T_0}\right)\right] \tag{22}$$

To obtain the best value of the activation energy for the different heating rates, the simulated data were graphed with temperature as a function of relative density and a fifth order polynomial fit was made over the range of $\rho = 0.6$ to $\rho = 0.9$. This polynomial was next used to obtain temperature versus relative densities at increments of 0.001. For each heating rate, the rhs of Equations (20), (21), and (22) were calculated for every given density. The absolute values of the differences between each heating rate, at a given density, were summed, and Solver in Excel was used to vary the activation energy until the sum reached a minimum. This process using Solver was performed for each of the three methods.

As can be seen in Table 2, the values for the activation energies obtained by all of the methods using the above minimization procedure show very good agreement as compared to the input value. Nearly all of the determined values are slightly below the input. Although numeric integration generally has lower values than the other two methods, all the numbers are so close that from a practical viewpoint, no technique is superior to the others.

Table 2. Activation energies determined by the three methods for different input values.

Input Q value (kJ/mol)	Numeric Integral Equation 20 (kJ/mol)	Analytical MSC Equation 21 (kJ/mol)	Lee and Beck Equation 22 (kJ/mol)
250	249.82	249.89	249.75
450	449.67	449.94	449.83
650	649.56	649.93	650.09

Figure 2 shows the MSCs for the three input values of the activation energy: Fig. 2A is for Q=250 kJ/mol, Fig. 2B is for Q=450 kJ/mol, and Fig. 2C is for Q=650 kJ/mol. For clarity, only 30 of the 300 calculated data points are shown for each curve. For each value of the activation energy, the general shape of the curves remains the same, although the value of $\ln[\Phi(\rho)]$ does become more negative as the activation energy increases. Of the three methods evaluated for determining the value of the integral, the numerical procedure (Equation (20)) and the analytic equation with $1/T$ removed from the integral, Equation (22), yield curves which are superimposed on each other. This suggests that the assumption is valid that the variation in $1/T$ is insignificant when compared to the variation of the exponential. Use of Equation (21) to determine the integral , however, showed significant deviation from the other two methods at low relative densities. This behavior is likely because the Lee and Beck approximation is used three times while generating the analytic expression, one of which further alters the results of an earlier approximation in that an approximation is squared and then added to itself. As would be predicted from the functional assumption of $(1+2RT/Q)$ being constant, the deviation is greatest at low temperature and reduces when $d\rho/dT$ is smaller. This deviation, however, seems to have no appreciable effect on the determined value of the activation energies, as seen in Table 2.

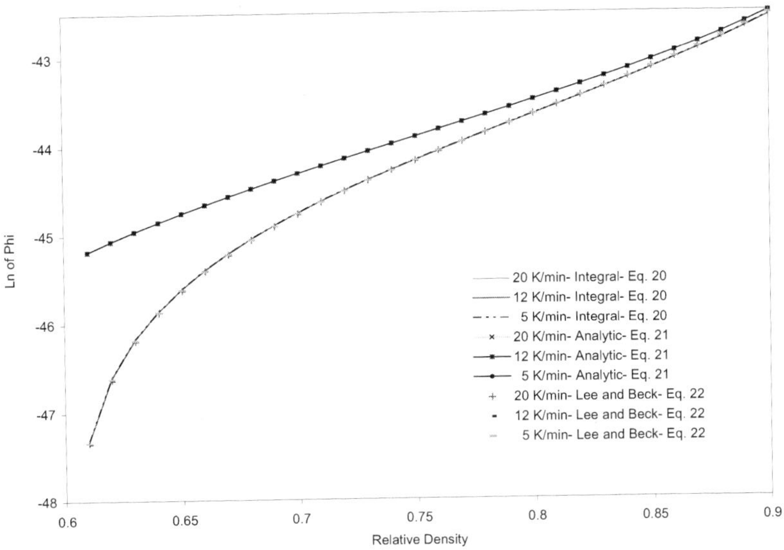
-43
-44
-45
-46
-47
-48
Ln of Phi
0.6
0.65
0.7
0.75
0.8
0.85
0.9
Relative Density
20 K/min- Integral- Eq. 20
12 K/min- Integral- Eq. 20
5 K/min- Integral- Eq. 20
20 K/min- Analytic- Eq. 21
12 K/min- Analytic- Eq. 21
5 K/min- Analytic- Eq. 21
20 K/min- Lee and Beck- Eq. 22
12 K/min- Lee and Beck- Eq. 22
5 K/min- Lee and Beck- Eq. 22

A)

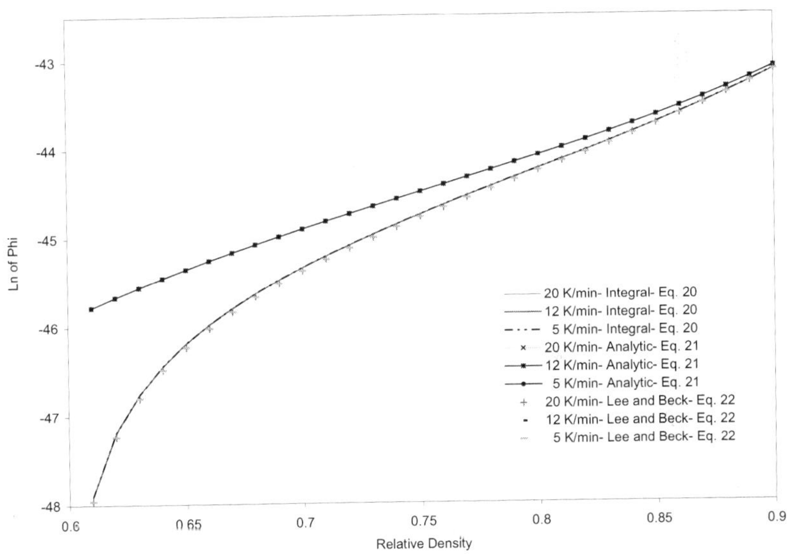
-43
-44
-45
-46
-47
-48
Ln of Phi
0.6
0.65
0.7
0.75
0.8
0.85
0.9
Relative Density
20 K/min- Integral- Eq. 20
12 K/min- Integral- Eq. 20
5 K/min- Integral- Eq. 20
20 K/min- Analytic- Eq. 21
12 K/min- Analytic- Eq. 21
5 K/min- Analytic- Eq. 21
20 K/min- Lee and Beck- Eq. 22
12 K/min- Lee and Beck- Eq. 22
5 K/min- Lee and Beck- Eq. 22

B)

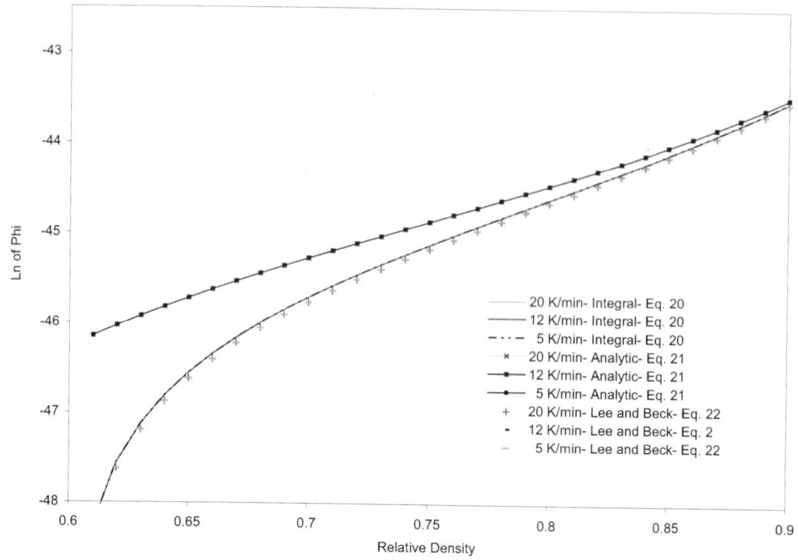

C)

Figure 2: MSCs for simulations created with Equation (19) at A) Q=250 kJ/mol, B) Q=450 kJ/mol, and C) Q=650 kJ/mol. In each figure, nine curves are shown representing the three heating rates with each analyzed set analyzed by the three methods. The results from Equations (20) and (22) are virtually indistinguishable from each other.

CONCLUSION

An approximation due to Lee and Beck has been used to obtain two analytical expressions for evaluating the integral that appears in the MSC method. The two analytical expressions where compared to trapezoidal numeric integration to obtain the activation energies for simulated density versus temperature data generated with activation energies from 250-650 kJ/mol. In general, both analytic expressions yielded values of the activation energy very close to the input values and to those obtained by numeric integration.

REFERENCES:

[1] H. Su, and D. L. Johnson, Master Sintering Curve: A Practical Approach to Sintering, *J. Am. Ceram. Soc.,* **79** [12] 3211-17 (1996).

[2] K. G. Ewsuk, and D. T. Ellerby, Analysis of Nanocrystalline and Microcrystalline ZnO Sintering Using Master Sintering Curves, *J. Am. Ceram. Soc,* **89** [6] 2003-09 (2006).

[3] S. Kiani, J. Pan, and J. A. Yeomans, A New Scheme of Finding the Master Sintering Curve, *J. Am. Ceram. Soc.,* **89** [11] 3393-3396 (2006).

[4] Y. Kinemuchi, and K. Watari, Dilatometer Analysis of Sintering Behavior of Nano-CeO_2 Particles, *J. Eur. Ceram. Soc.,* **28** 2019-2024 (2008).

[5] M. H. Teng, Y. C. Lai, and Y. T. Chen, A Computer Program of Master Sintering Curve Model to Accurately Predict Sintering Results, *West. Pac. Earth Sci.*, **2** [2] 171-180 (2002).

[6] A. Mohanram, G. L. Messing, and D. J. Green, Densification and Sintering Viscosity of Low-Temperature Co-Fired Ceramics, *J. Am. Ceram. Soc.,* **88** [10] 2681-89 (2005).

[7] T. V. Lee, and S. R. Beck, A New Integral Approximation Formula for Kinetic Analysis of Nonisothermal TGA Data, *AIChE J.,* **30** [3] 517-519 (1984).

[8] J. D. Hansen, R. P. Rusin, M. Teng, and D. L. Johnson, Combined-Stage Sintering Model, *J. Am. Ceram. Soc.,* **75** [5] 1129-35 (1992).

SURFACE ANALYSIS OF NANO-STRUCTURED CARBON NITRIDE FILMS FOR MICROSENSORS

Choong W. Chang[1], Ju N. Kim[1], Yoen H. Jeong[1], Young J. Seo[1], S. Chowdhury[2], Sung P. Lee[1]

[1]Department of Electronic Engineering, Kyungnam University, Masan, Korea
[2]Intel Corporation, Hilsboro, OR 97124, USA

ABSTRACT

Nano-structured carbon-nitride films have been deposited by rf magnetron sputtering system for application of micro humidity sensors. The surface analysis were investigated by FTIR, SEM, AFM and EDS. The film has a uniform nano-structured surface morphology and the average grain size is about 20 nm. FTIR spectrum indicates an α-C_3N_4 peak. The band at ~2200 cm^{-1} is assigned to approximately stretching vibration of the ∈N bond. The C=N stretching band (sp ^{2} line) appears at between 1600~1700 cm^{-1} depending on the level of nitrogen incorporation in the film. In the interdigitated-electrodes (IDE) type, the resistance of the film changed from 8283.30 kΩ to 827.63 kΩ. In the sandwich type, the resistance decreased from 577.41 kΩ to 24.95 kΩ, however, the capacitance increased from 504.31 nF to 1.30 nF in the relative humidity range of 10 to 90 %.

INTRODUCTION

There has been much interest in the deposition of crystalline carbon nitride films since the theoretical properties of such material were calculated by Liu and Cohen[1,2]. Carbon nitride films(especially β-C_3N_4) were predicted to be a superhard material, comparable to or greater than that of diamond, due to a short bond length(1.47 A) and low ionicity(~ 7 %). A number of groups have attempted to synthesize crystalline β-C_3N_4 in a variety of techniques such as reactive sputtering[3,4], CVD[5], laser ablation[6], ion beam assisted deposition[7]. Despite a large number of efforts to synthesize this material over the ten years, most of the obtained films were almost amorphous CN_x and were not as good as predicted.

One of the most difficult problems is to avoid hydrogen attacks not only during the films were being deposited but also when they were exposed in the atmosphere. Hydrogen causes excessive stress that is high enough to swell the adhesion between the substrate and the film due to the presence of moisture in the chamber wall as well as out of the chamber. This hydrogen attack can easily breaks or changes the C≡N and C=N bonds to C-H and N-H bond formation[8,9]. This is strongly undesirable behavior for the composition of the crystalline β-C_3N_4. but if one can use hydrogen attractive characteristics due to some hydrogen defects which were made intentionally and weak bonded, CN_x could be a new material for humidity sensor applications.

In this work, the properties of the bonding structure and composition rate of as-deposited carbon nitride films has been described as a function of their nitrogen content and the heat treatment. We report the change of electrical properties of nanostructured CN_x thin film with ambient humidity. The films are formed by the reactive-magnetron sputtering method, and photolithography and lift-off technique are used for electrode formation.

EXPERIMENTAL

Carbon nitride films were grown on polished n-type (100) silicon substrates by reactive magnetron sputtering system with DC bias using facing targets as shown schematically in Fig. 1. The use of a Nd-Fe-B magnet ensures a very high flux density between the substrate and target region. The substrates were ultrasonically cleaned in acetone and methanol. The sputtering targets were graphite of 99.95% purity and the sputter gas was an Ar/N_2 mixture of variable composition. The target was pre-sputtered for 20 min prior to each deposition to remove surface contamination. Then the nitrogen (99.999 %) gas and Ar (99.999 %) gas were introduced into the sputtering chamber through a mass-flow controller (MFC) until the required pressure was obtained. The nitrogen contents of the deposited films were varied by controlling the nitrogen partial pressure in the sputtering gas. To avoid the hydrogen contamination from water molecules on the chamber wall, the chamber was heated to 200 ℃ with the resistively-heated coil. The deposition conditions of the carbon nitride film are summarized in Table 1.

Fourier transform infrared (FTIR) absorption spectra were included for revealing the bonding structures of deposited films with Research I Series (Mattson, UK). For the morphology of the films, Scanning electron microscopy (SEM) photograph was used with ABT-32 TOPCON (Japan), and scanning probe microscopy (SPM) with a contact atomic force microscopy(AFM) mode from ThermoMicroscops (USA) was used to measure the surface roughness and to observe the surface topography of the films. The films were conductive enough to ignore the charging effect. Energy Dispersive X-ray Spectroscopy (EDS, Kevex Sigma MS2, USA) with 138 eV resolution are also used to confirm the quantitative study of the carbon nitride. Heat treatment was carried out at temperatures up to 450 ℃ for 30 min.

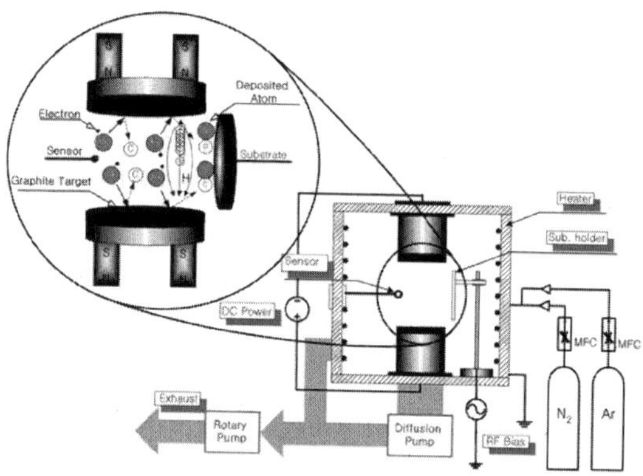

Fig. 1. Schematic diagram of a reactive RF magnetron sputtering system with DC bias.

Table 1. Deposition conditions of carbon nitride film

Deposition Condition	
Target	Graphite
RF power (W)	200, 300, 400
Substrate bias voltage (V)	DC -60
N_2/Ar ratio	0/10, 3/7, 5/5, 7/3, 10/0
Base pressure (Torr)	7×10^{-5}
Working pressure (Torr)	5×10^{-2}
Temperature (°C)	200
Deposition time (min)	60

Two conventional types for humidity sensor were designed for measurement of the humidity characteristics as shown in Fig. 2. Lift-off technique was applied for interdigitated Au electrodes (Fig. 2(b)); 20 μm thick, 0.4 mm wide and 0.2 mm distance between fingers. The photo resist (TSMR-8900D, TOK) and an additive to improve an adhesion were spin-coated on the initial cleaned Si substrate. For electrodes, Au which was evaporated with thermal evaporator was removed by the lift-off technique using acetone. Then, CN_x film was deposited on the substrate. The similar process and method were applied for the sample of capacitance type (Fig. 2(a)). The CN_x film was deposited on the lower Al electrode which was evaporated on the Si substrate by thermal evaporator. To improve hysteresis, small meshes are patterned in the upper electrode (1 mm x 1 mm) by lift-off technique.

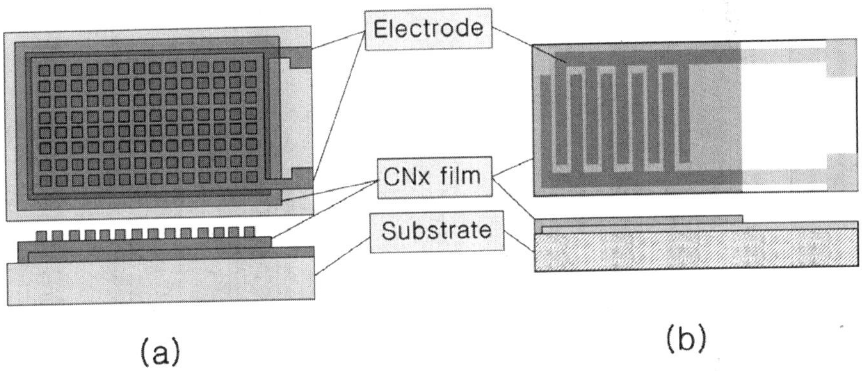

Fig. 2. Schematic view of two types of sensor; capacitance type with a top electrode of small meshes (a) and resistance type with interdigitated electrode (b).

RESULTS AND DISCUSSION

In order to understand the bonding state of the CN_x films, FTIR analysis was performed. The carbon nitride films were deposited on the quartz substrate with RF power of 200 W, bias of - 60 V_{DC} and time for 60 min. Fig. 3 shows the FTIR absorption spectra of three samples with different nitrogen ratio and bare quartz substrate for the reference. We can find that there are several peaks of α-C_3N_4 and β-C_3N_4 in Fig. 3. The peaks at 1011 cm^{-1} and 1257 cm^{-1} are attributed to α-C_3N_4 absorption bands and the weak peak at 1529 cm^{-1} is β-C_3N_4 absorption band. The band between 1500 cm^{-1} and 1800 cm^{-1} is due to sp^2 carbon atoms bonded to nitrogen atoms which form the main network of the materials [10]. The weak peak of an adjacent to 2200 cm^{-1} is due to the C≡N stretching mode. Two peaks centered at 2900 cm^{-1} and 3300 cm^{-1} due to the hydrogen influence are CH and NH_2 stretching mode. These results show that the CN_x film has some of C-H and N-H bonds and the film can easily absorb and desorb the OH radicals. If the film has some hydrogen defect, the film can offer the binding site with water vapor. Hydrogen can make C-H and N-H bond easily from C≡N and C=N bonds. From this result, we concluded that these effects can prepare CN_x film having a good dependence on the relative humidity.

Surface morphology of the CN_x film was investigated by SEM. Fig. 4 gives typical surface morphology of the film deposited with 50 % N_2 ratio and 200 W RF power for 1hr. The film was composed of nano-structured grains(ca. 20 ~ 40 nm) which have a high density.

The surface roughness also investigated by AFM (Fig. 5). The topography of the film surface was composed of very fine grains with an rms roughness of 1.2188 nm.

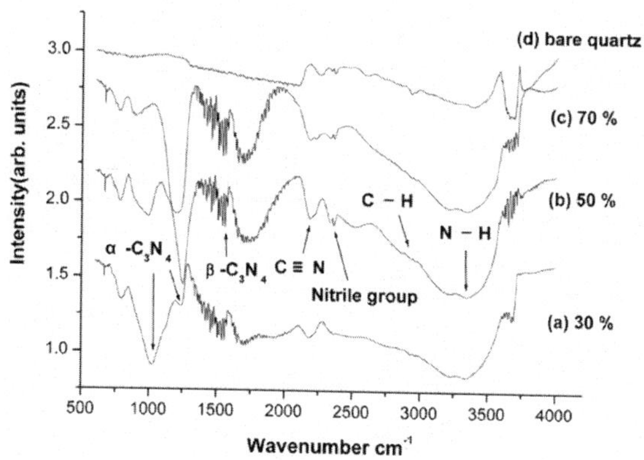

Fig. 3. FTIR spectra of CN_x films as function of relative nitrogen ratio ($N_2/(N_2+Ar)$).

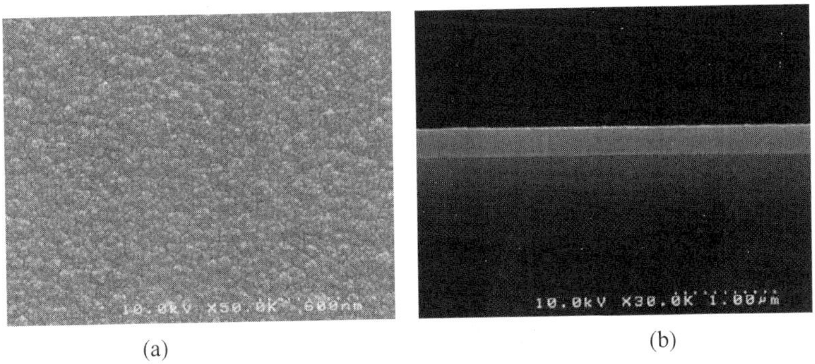

Fig. 4. SEM photograph of the CN_x film on the Si wafer with 50 % N_2 ratio and 200 W RF power for 1hr; (a) surface and (b) cross section.

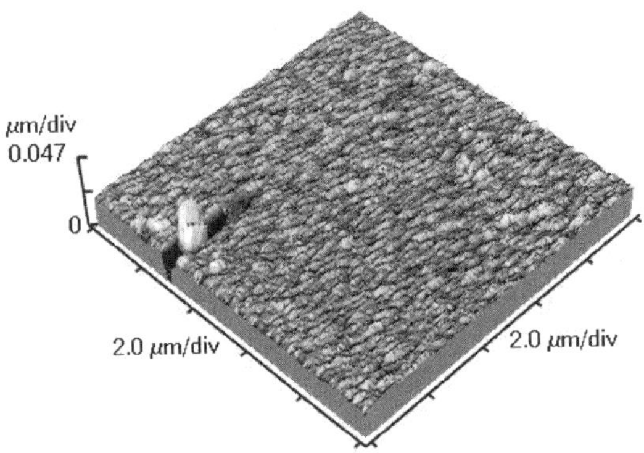

Fig. 5. AFM image of the CNx film deposited on the Si wafer with 50 % N_2 ratio and 200 W RF power for 1hr.

The composition ratio of deposited films is also conformed by EDS which can measure deeper surface than XPS or AES. Fig. 6 shows the normalized EDS peaks of the CN_x samples. The carbon/nitrogen ratio is 51.97/36.11 at 70% N_2 gas. Unwanted nitrogen appears in 0% N_2 gas. It is supported that carbon nitride around carbon target is sputtered during next sputtering process nevertheless the pre-sputtering. The detail of component is shown in table 2. As the N_2 sputtering gas ratio increases, nitrogen incorporation in the film also increases under the condition that the nitrogen gas ratio is lower than 70%. However, when N_2 ratio is higher than 70%, the nitrogen incorporation of the film is even decreased. This is probably due to increased sputtering rate of C species from the target as the steady state N concentration increases on the target surface. Energetic or neutral Ar species probably enhance the mobility of nitrogen species and increase N sticking at the growth surface. There is indeed an effect of Ar mixture on the nitrogen incorporation in the film. When N_2/Ar ratio is 30~70%, sputtering rate of carbon and nitrogen sticking effect is maximum at the growth surface. If Ar increases further in the sputtering gas mixture, two things can happen [11], (1) the sputtering rate of carbon increases with respect to ionized nitrogen species, thus the film contains less nitrogen: (2) chemically enhanced preferential sputtering of nitrogen from the film surface can occur by the energetic Ar species. Increase in Ar gas in the sputter gas mixture may also disrupt the film structure. The momentum of the Ar^+ ions is higher than that of either N_2^+ or N^+ ions, and then the increased momentum transfer into the growing film causes disruption of the bonding structure of the film producing amorphous material. We can find that if N_2 increases more than 70% in the sputtering gas mixture, sputtering yield and N sticking effect are decreased due to deficiency of energetic Ar species. It can be seen that with higher than 70% N_2 sputtered film, nitrogen incorporation is lower than the 50% N_2 gas sputtered film. A small amount of oxygen (less than 6.23 atomic%) is due to contamination from the chamber wall, and Si peak is caused by the silicon wafer substrate.

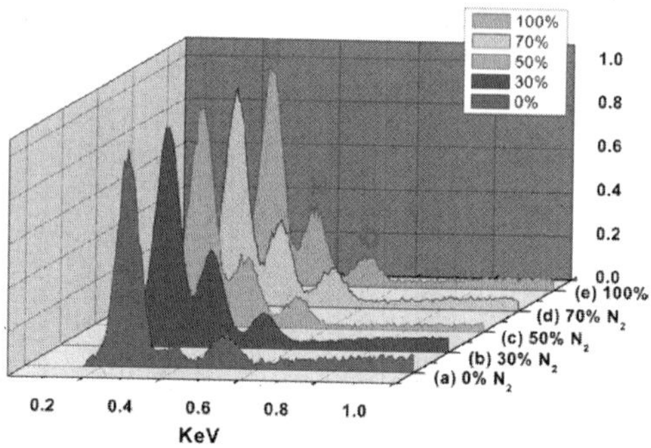

Fig. 6. Normalized EDS spectra of CN_x films with different N_2 ratio.

Table 2. Atomic ratio of CN_x films by EDS results.

Element \ N_2 ratio	0	30	50	70	100
C	64.93	55.14	71.68	51.97	69.41
N	4.7	31.59	23.63	36.11	24.44
O	2.64	4.60	6.23	3.67	4.81
Si	25.9	8.66	5.69	1.03	1.34

Fig. 7 shows the resistance dependence on the relative humidity of CN_x films with Au-IDE at 50 % N_2 ratio and 200W RF power for 1 hr (ca. 1.5 μm thickness). The sensor was connected to LCR meter at 1 V and 100 kHz. The resistance of the CN_x film decreases as increasing the relative humidity and especially, the slope decreases rapidly over the 65 %RH. The resistance of the film changes from 8.2833 MΩ to 827.63 kΩ in the range of 10 and 90 %RH. At the high humidity region, however, the film operates very stable and has no the fluctuation of resistance. This result shows that the sensitivity to humidity is better at high humidity region than low humidity one. It is shown that CN_x films can be used for dew point sensors.

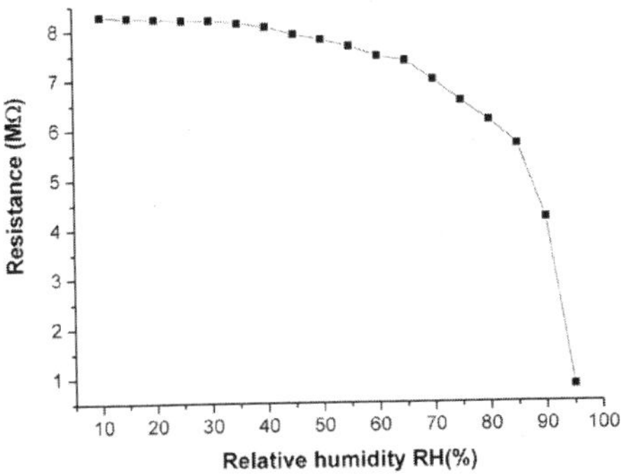

Fig. 7. Resistance dependence on the relative humidity for the CN_x film; N_2 ratio: 50 %, power: 200 W, time: 1 hr, operating temperature: 30 °C, frequency: 100 kHz and applied voltage: 1 V.

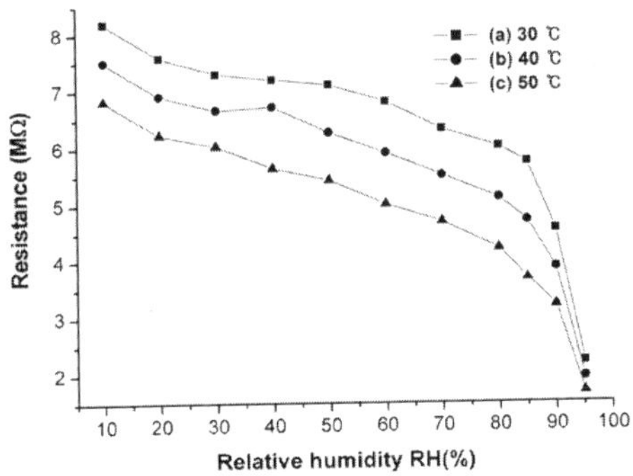

Fig. 8. Resistance dependence on the relative humidity for the CN_x film (70 % N_2 ratio, 200 W, 1 hr) as a function of chamber temperatures.

Fig. 8 shows the temperature dependence of the CN_x film which has the same structure and measurement condition as the sample of Fig. 7, except N_2 ratio of 70 %. The measurement chamber temperature was fixed at 30 °C (Fig. 8(a)), 40 °C (Fig. 8(b)) and 50 °C (Fig. 8(c)), respectively. As shown in Fig. 8, the film resistance decreases as increasing the relative humidity in the all temperature range. The sensitivity in low humidity region of the film which is deposited at 70 % N_2 ratio is better than the film at 50 % N_2 ratio compared with Fig. 7, that is, the slope is more linear. As increasing the operating temperature, the resistance of film decreases at same humidity, however the slope was not changed. This result reveals that the change of characteristics to the different operating temperatures can easily be compensated by the simple computer programming in the microprocessor or analog circuit system.

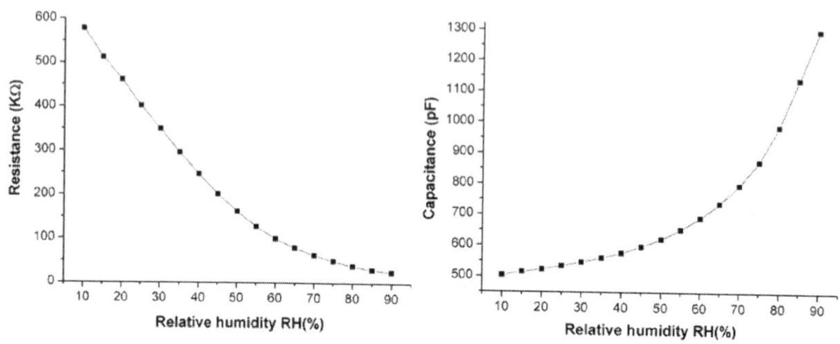

(a) (b)

Fig. 9. Resistance dependence(a) and capacitance dependence(b) on the relative humidity for the CN_x film.

Fig. 9 shows the electrical characteristics of the CN_x films with a capacitance type. The film was deposited on the lower electrode with 50 % N_2 ratio, 200 W RF power and for 1 hr. For the measurement, LCR meter has been used with 10 kHz and 1 V bias. As shown in Fig. 9, resistance and capacitance both change to a relative humidity in the humidity range between 10 and 90 %RH. The resistance of the film decreases from 577.41kΩ to 24.949kΩ (Fig. 9(a)), but the capacitance increases from 504.31nF to 1.2989nF (Fig. 9(b)) at 30 °C. The resistance response is more linear than the capacitance one, when the Y-axis is converted to the logarithm scale. It is considered that this capacitive type sample can operate at full range of relative humidity.

CONCLUSION

Carbon nitride film for a new Si-based humidity sensor material has been deposited by RF magnetron sputtering system. With increasing relative humidity, the resistance of CN_x films decreased and the capacitance increased. The film which was deposited with 50 % N_2 ratio, 200 W RF power for 1 hr revealed a good humidity dependence with wide range between 10 and 90 %RH. According to this study, CN_x ilms can be used new humidity-sensing material for dew compensation sensor or wide range relative humidity sensor, especially MEMS based or Si based nano-structured micro-humidity sensors.

ACKNOWLEDGMENT

This research was financially supported by the Ministry of Education, Science Technology (MEST) and Korea Industrial Technology Foundation (KOTEF) through the Human Resource Training Project for Regional Innovation

REFERENCES

[1] M. L. Cohen, Phys. Rev. B, 32, 7988 (1985).
[2] A. Y. Liu and M. L. Cohen, Science, 245, 841 (1989).
[3] K.M. Yu, M.L. Cohen, E.E. Haller, W.L. Hansen, A.Y. Liu and I.C. Wu, Phys. Rev. B: Condens. Matter, B 49, 5034 (1994).
[4] S. P. Lee and J. B. Kang, Microchemical Journal, 70, 239 (2001).
[5] H. Han and B. J. Feldman, Solid State Commun., 65, 921 (1988).
[6] P. Gonzalez, R. Soto, E.G. Parada, X. Redondas, S. Chiussi, J. Serra, J. Pou, B. Leon and M. Perez-Amor, Appl. Surf. Sci., 109-110, 380 (1997).
[7] K. J. Boyd, D. Marton, S.S. Todorov, A.H. Al-Bayati, J. Kulik, R.A. Zuhr and J. W. Rabalais, J. Vac. Sci. Technol., A 13, 2110 (1995).
[8] S. P. Lee and J. B. Kang, S. Chowdhury, J. Korean Phys. Soc., 39, S1 (2001).
[9] H. J. Schotzau and S. Veprek, Appl. Phys., 7, 27 (1975).
[10] C. Popov, M.F. Plass, R. Kassing and W. Kulisch, Thin Solid Films, 333-356, 406 (1999).
[11] W. Zheng, T. Ding, I. Ivanov and J. E. Sundgren, J. Mater. Sci. Technol., 13, 154-158 (1997).

GAS PERMEABILITY IN NANOPOROUS SUBSTRATES

S. J. Lombardo*, J.W. Yun, and S. Patel
Department of Chemical Engineering, University of Missouri, Columbia, MO 65211, USA

*Department of Mechanical and Aerospace Engineering, University of Missouri, Columbia, MO 65211, USA

ABSTRACT

To reduce component size, nanosized powders are being used more frequently in the fabrication of multilayer ceramic capacitors. As a consequence of the small particle size, the gas permeability can be greatly reduced, which makes binder removal by thermal methods a more difficult process. In this paper, we show how the permeability can be determined in the nanopore size range from the combined outcome of both experiments and modeling. This approach utilizes measurements of the flux through a substrate to determine the pore size, which is then used to determine the relative contributions from Knudsen, slip, and laminar flow. This technique is applied to a membrane material with specified, narrow pore size.

INTRODUCTION

The permeability [1-6] of gases through porous media is a property that may influence a number of other processing operations in the fabrication of ceramic components, including binder removal [7-12] from green ceramic bodies. As such, it is important to have techniques available to characterize this property. In addition, as the processing of nanosized powders becomes more prevalent, the permeability of green bodies prepared from such powders becomes lower, which may make the processing of such powders more difficult.

An additional complexity is that as the particle size gets smaller, which in turn leads to smaller pore sizes, the flow mechanism in the pore space may become more complicated as Knudsen or slip flow or both may occur in addition to laminar (Poiseuille) flow [13-19]. To determine which of these flow mechanisms is the most dominant, it is necessary to know the pore size of the porous medium as well as the mean free path of the molecules in the gas phase. In general at low pressures, as the pore size increases from 0.1 μm to 1 μm, the flow transitions from slip-dominated to Poiseuille-dominated flow. Although techniques such as mercury porosimetry can be used to characterize the pore size, the resulting pore size is relevant for characterizing liquid-type permeability. For gas permeability, it is preferable to have a techniques based on gas-phase probe molecules.

Fortunately, such a technique was outlined many years ago [16], and this method is based on measuring the flux at different pressure drops in a flow regime where Knudsen flow can be neglected. Then, a graphical procedure is used to determine the characteristic pore size. In this paper, we use such measurements of the gas flux to determine the permeability and pore size of membrane substrates with a very uniform and narrow pore size. We then use this approach, as applied by us earlier [20-21], in order to assess the error and reproducibility of such measurements.

EXPERIMENTAL

In this work, anodized aluminum substrates (Whatman International Ltd, Anodisc 25, Material No. : +H718680960021K) were used and are of nominally of 0.02 μm pore size. Figure 1 shows, however, that the actual average pore size is much larger, and is closer to 0.2 μm; the pores also appear fairly straight.

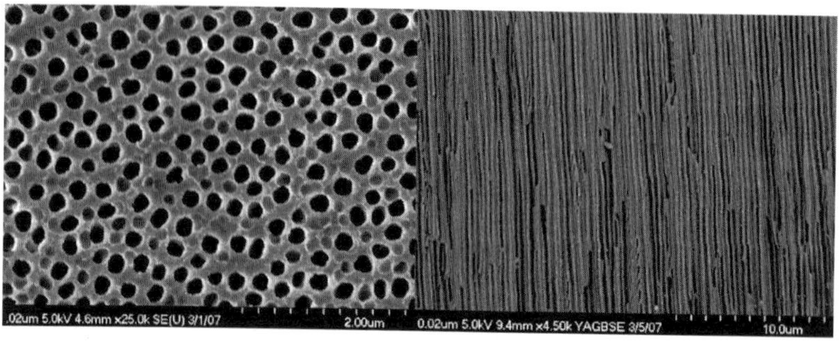

Figure 1 Scanning electron micrograph images of a substrate showing top view (left) and cross section (right).

To measure the permeability of the substrates, either one or two samples were placed in a membrane holder (Millipore, Billerica, MA; see Fig. 2) and sealed with a polymeric o-ring. A pressure regulator and valve were used to set an upstream pressure of nitrogen, and the downstream pressure was approximately atmospheric pressure. The volumetric flow rate, V_f, was determined by a bubble meter, and the pressure drop across the substrate was measured with a U-tube manometer containing water. The molar flow rate, n, was then determined by

$$n = \frac{P_f V_f}{R T_f} \quad (1)$$

where P is the pressure, R is the gas constant, T is the temperature, and the subscript f denotes the conditions at the flow meter. The molar flux, N, through the substrate area, A, can then be calculated by

$$N = n/A = u_o \rho \quad (2)$$

where u_o is the superficial velocity and the molar density is given by $\rho = P/RT$.

The procedure to determine the characteristic pore size is as follows [16]. Measurements of the gas flux versus a range of pressure drops are made, and the normalized flux, $N/\Delta P$, is then graphed versus the average pressure, P_{ave}, across the sample. For experiments conducted at pressures for which Knudsen flow can be neglected, linear behavior may rise as given by

$$\frac{N}{\Delta P} = AP_{ave} + B \quad (3)$$

where A is the slope and B is the intercept. The ratio A/B can then be used to determine the pore size as

$$\frac{A}{B} = \frac{3\bar{r}}{4\pi\mu\bar{v}} \qquad 4$$

where μ is the gas viscosity and $\bar{v} = (8RT/\pi M)^{1/2}$ is the gas velocity from the kinetic theory of gases. The characteristic pore diameter is then given as $D = 2\bar{r}$.

For flow in pores which is mainly governed by laminar flow, the permeability can be determined from Darcy's Law for a compressible fluid as

$$\kappa = -2RTN\mu \frac{L}{P_2^2 - P_1^2} \qquad 5$$

where L is the thickness of the substrate.

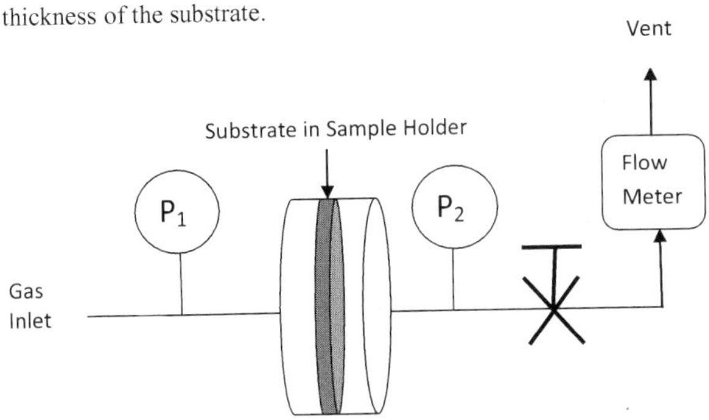

Figure 2 Schematic of the flow apparatus used to determine the permeability and characteristic pore size.

RESULTS AND DISCUSSION

Flux experiments were conducted though two individual membrane samples, and then these same two membranes were tested when combined into a single stack. Figure 3 shows the normalized flux from two tests versus the average pressure for Sample 1. With increasing average pressure, the normalized flux first decreases and then goes through a shallow minimum before increasing in a linear manner. Both samples exhibit qualitatively similar behavior, but the curves are shifted by 3-4% relative to each other. The origin of the minimum is unknown.

Sample 2 was tested twice in a similar manner, and Fig. 4 shows that for the two tests the results are qualitatively similarly to each other and also to the results for Sample 1 in Fig. 3. From a quantitative perspective, the two trials for Sample 2 are within 2% of each other, and within 6% of the results for Sample 1.

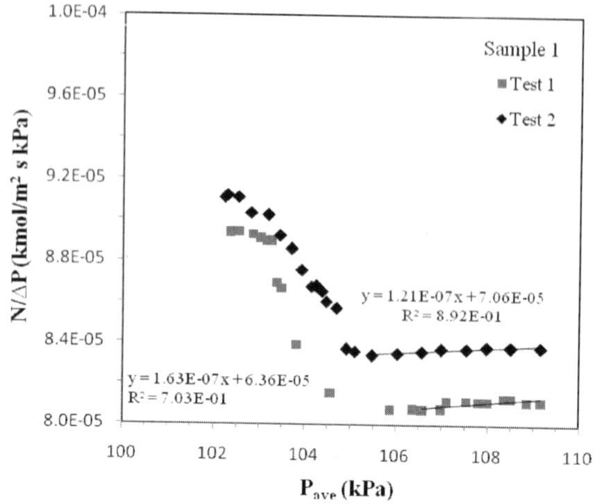

Figure 3 Normalized flux versus average pressure for two tests for Sample 1. The best fit lines for the linear region of behavior are also shown.

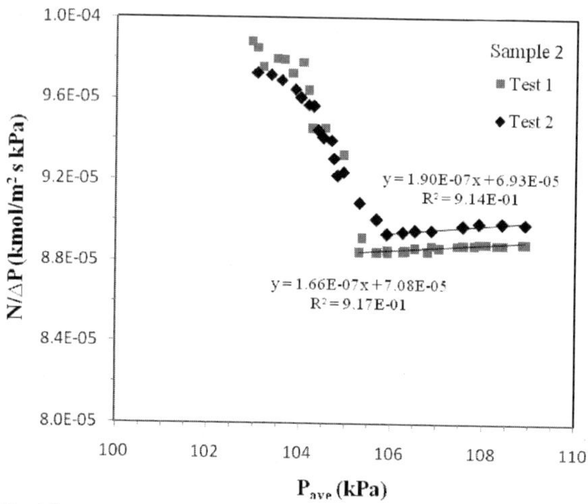

Figure 4 Normalized flux versus average pressure for two tests for Sample 2. The best fit lines for the linear region of behavior are also shown.

Figure 5 shows the normalized flux behavior when Sample 1 and Sample 2 are combined into a single stack (denoted as Sample 1+2) so that the overall thickness of the stack is twice that of an individual membrane. The two tests for Sample 1+2 agree well with each other, and qualitatively exhibit the same behavior as for the individual Sample 1 and Sample 2. For the same pressure drop, however, the normalized flux is approximately half that seen in Figs. 3 and 4, and this arises from the greater resistance to flow because the stacked samples are twice the thickness.

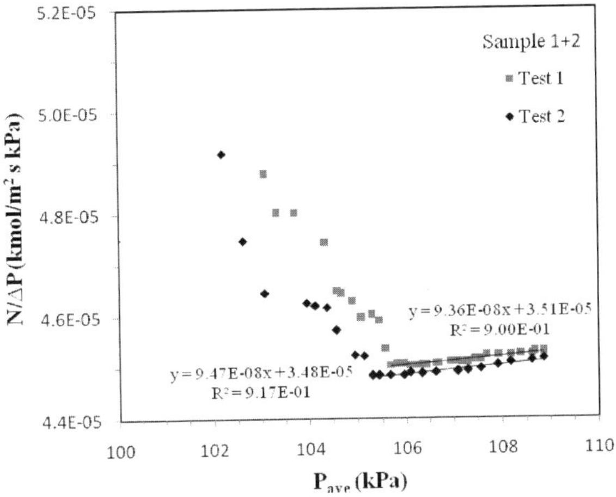

Figure 5 Normalized flux versus average pressure for two tests for Sample 1+2 tested in a stacked configuration. The best fit lines for the linear region of behavior are also shown.

The flux values in Figs. 3-5 were then used with Eq. 5 to determine the permeability of the samples. Figure 6 shows that for all of the samples, the permeability initially decreases with increasing inlet pressure and then appears to either plateau at a constant value or to decrease weakly with further increases in pressure. In general, however, the permeability values for the multiple tests for each sample agree quite well with each other, and additionally, all of the permeability values only vary by 20%. The differences may arise from either errors in testing or from small differences between each sample. The decrease in the permeability may arise because as the pressure increases, the mean free path decreases, and thus the contribution from slip flow is decreasing. Under these circumstances, Eq. 5 is not strictly valid, as it is derived for the case of laminar flow only. Nevertheless, for engineering purposes, analysis of the flux data with Eq. 5 seems to be reasonably justified in light of the small variations in the permeability seen in Fig. 6.

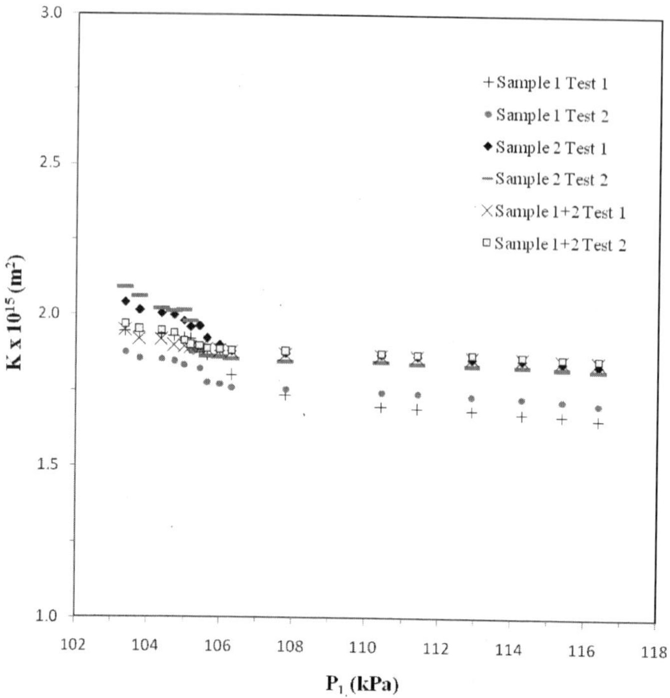

Figure 6 Gas permeability versus upstream pressure for two testes each for Sample 1, Sample 2, and Sample 1+2.

The results in Figs. 3-6 indicate that a fairly high degree of reproducibility exists between the individual samples and the re-tests, especially in light of the fact that permeability values can vary over a wide range and that models to describe the permeability often require one or more adjustable parameters. To lend some insight into the value of the underlying quantities that comprise the permeability, Fig. 7 shows values of the molar flux, superficial velocity, gas density, and permeability for test 2 for Sample 1+2 as a function of inlet pressure. Both the superficial velocity, which is calculated at the center of the sample, and molar flux increase with increasing inlet pressure, and hence pressure drop. The gas density, however, decreases slightly with increasing inlet pressure, as expected from the ideal gas law. The combined influence of these variations, however, leads to a nearly constant value of the permeability, which is expected if the permeability is a property of the porous medium, independent of the flow conditions. This expectation for the constancy of the permeability is predicated on the idea, however, that the flow mechanism is not changing.

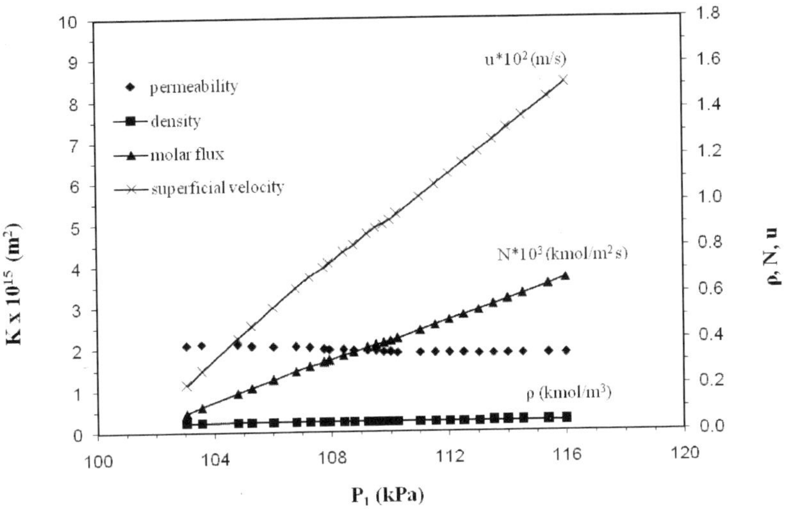

Figure 7 Gas permeability (left-hand axis) and gas density, gas flux, and gas superficial velocity (right-hand axis) versus upstream pressure for test 2 for Sample 1+2.

The normalized flux data in Figs. 3-5 can also be used to determine the pores size using Eqs. 3 and 4. This is accomplished by fitting the linear region of each curve and then using regression analysis to obtain the slope and the intercept. The regression lines in the Figs. 3 and 4 indicate that reasonable regions of linearity can be obtained, but the slopes and intercepts display a fair amount of variability between each sample and each test, up to 25%, as seen in Table 1. The values of the pore size determined in this manner are also in Table 1, and they generally fall within a narrow range of values of 0.12-0.19 μm. For this range of pore size, both laminar and slip flow contribute substantially to the total flux, and thus Eq. 5 is only approximately valid for representing the gas permeability.

Table 1 also contains the 95% confidence intervals for the slope and the intercept for each test, and it is seen that the slopes have a large degree of uncertainty between them. This uncertainty then manifests itself in a wide range for the confidence intervals for the pore sizes. Table 1 also summarizes the average permeability and standard deviation for each sample. In general, the permeability for each sample is very consistent, although sample to sample variations do occur.

Table 1 Values for the constants A and B (including 95% confidence intervals), pore diameter (including 95% confidence intervals), and average permeability (including standard deviations) for different substrates and tests.

Sample	A x 10^{-7} (kmol/m^2 s kPa^2)	B x 10^{-5} (kmol/m^2 s kPa)	D (μm)	K x 10^{15} (m^2)
Sample 1				
Test 1	1.63 ± 1.47	6.36 ± 1.58	0.18 ± 0.16	1.81 ± 0.11
Test 2	1.21 ± 0.42	7.06 ± 0.45	0.12 ± 0.04	1.79 ± 0.06
Sample 2				
Test 1	1.66 ± 0.40	7.09 ± 0.44	0.16 ± 0.04	1.99 ± 0.12
Test 2	1.90 ± 0.58	6.93 ± 0.63	0.19 ± 0.06	1.98 ± 0.09
Sample 1+2				
Test 1	0.94 ± 0.22	3.51 ± 0.23	0.19 ± 0.05	1.95 ± 0.09
Test 2	0.95 ± 0.18	3.48 ± 0.19	0.19 ± 0.04	1.92 ± 0.07

CONCLUSIONS

The permeability and pore size of substrates of narrow pore size and pore size distribution were analyzed by measurements of the gas flux through membrane samples. In general, the flux behavior was very similar between tests and samples, which lead to consistent values for the permeability. The determination of the pore size, however, was slightly less consistent, but the average pore size was in good agreement with the visual appearance of the pore as determined by microscopy. The location of the minimum in the behavior of the normalized flux versus average pressure, however, remains to be explained.

REFERENCES

1. A.E. Scheidegger, *The Physics of Flow Through Porous Media,* University of Toronto Press, Toronto, 1974.
2. F. A. L. Dullien, *Porous Media: Fluid Transport and Pore Structure,* Academic Press, New York, 1979.
3. J. Bear, and J.-M. Buchlin, *Modelling and Applications of Transport Phenomena in Porous Media*, Kluwer Academic Publishers, Dordrecht, 1991.
4. M. Sahimi, *Flow and Transport in Porous Media and Fractured Rock*, VCH, Weinheim, 1995.
5. P.C. Carman, *Flow and Transport Through Porous Media*, Academic Press, New York, 1956.
6. H. David, "A Review of Terminology Pertaining to Darcy's Law and Flow through Porous Media," *Journal of Porous Media*, **6** [2] 87-97 (2003).
7. R. M. German, "Theory of Thermal Debinding," *Int. J. Powder Metall.*, **23** [4] 237-245 (1987).
8. J. A. Lewis, "Binder Removal From Ceramics," *Annual Rev. Mater. Sci.*, **27**, 147-173 (1997).
9. G. Y. Stangle, and I. A. Aksay, "Simultaneous Momentum, Heat and Mass Transfer With Chemical Reaction in a Disordered Porous Medium: Application to Binder Removal from a Ceramic Green Body," *Chem. Eng. Sci.*, **45**, 1719-1731 (1990).
10. D-S. Tsai, "Pressure Buildup and Internal Stresses During Binder Burnout: Numerical Analysis ," *AIChE J.*, **37** [4] 547-554 (1991).
11. J. H. Song, M. J. Edirisinghe, J. R. G. Evans, and E. H. Twizell, "Modeling the Effect of Gas Transport on the Formation of Defects during Thermolysis of Powder Moldings," *J. Mater. Res.*, **11**, 830-840 (1996).
12. S. J. Lombardo and Z. C. Feng, "Analytic Method for the Minimum Time for Binder Removal

from Three-Dimensional Porous Green Bodies," *J. of Mat. Res.*, **18,** 2717-2723 (2003).
13. L. J. Klinkenberg, "The Permeability of Porous Media to Liquids and Gases," *Drill. Prod. Prac. API*, 200-213 (1941).
14. G. P. Brown, A. DiNardo, G. K. Cheng, and T. K. Sherwood, "The Flow of Gases in Pipes at Low Pressures," *J. Appl. Phys.*, **17**, 802-813 (1946).
15. D. S. Scott and F. A. L. Dullien, "Diffusion of Ideal Gases in Capillaries and Porous Solids," *AIChE J.*, **8** [1] 113-117 (1962).
16. D. S. Scott and F. A. L. Dullien, "The Flow of Rarefied Gases," *AIChE J.*, **8** [3] 293-297 (1962).
17. N. Wakao and J. M. Smith, "Diffusion in Catalyst Pellets," *Chem. Eng. Sci.*, **17**, 825-834 (1962).
18. N. Wakao, S. Ontani, and J. M. Smith, "Significance of Pressure Gradients in Porous Materials: Part 1. Diffusion and Flow in Fine Capillaries," *AIChE J.*, **11** [3] 435-439 (1965).
19. S. Ontani, N. Wakao, and J. M. Smith, "Significance of Pressure Gradients in Porous Materials: Part 2. Diffusion and Flow in Porous Catalysts," *AIChE J.*, **11** [3] 439-445 (1965).
20. J. W. Yun and S. J. Lombardo, "Permeability of Green Tapes as a Function of Binder Loading," *J. Am. Ceram. Soc.*, **90** [2] 456- 461 (2007).
21. J. W. Yun and S. J. Lombardo, "Permeability of Laminated Green Tapes as a Function of Binder Loading," *J. Am. Ceram. Soc.*, **91** [5] 1553-1558 (2008).

Properties and Applications

TEXTURING OF PMN-PT CERAMICS VIA TEMPLATED GRAIN GROWTH (TGG): ISSUES AND PERSPECTIVES

Mohammad E. Ebrahimi
SorenTec
Toronto, Ontario, Canada

ABSTRACT

The issues in texturing of PMN-PT ceramics by Templated Grain Growth (TGG) are discussed. TGG is a technique to make grain oriented ceramics with high anisotropy properties in lower cost. In recent years, extensive researches are focused on obtaining higher microstructural orientation. High d_{33} (1100-1600 pC/N) in low field and high strain (>0.3% at 50 kV/cm) were reported. Nevertheless, the technique still faces some problems in achieving high and reliable piezoelectric properties.

High aspect ratio strontium titanate (ST) and barium titanate (BT) are synthesized and incorporated in PMN-35PT matrix. High Lotgering factors (~70%) were obtained. Synthesis of calcium titanate (CT) was also studied. Principles and developments of viable high isometric perovskite templates and the perspective of the TGG technique for texturing of PMN-PT are presented. Finally, fluctuations in piezoelectric properties of sintered samples are discussed in terms of compositional and microstructural aspects.

INTRODUCTION

In recent years, controlled texture development in polycrystalline materials has been as an interesting topic in ceramic processing, because it results in improved electrical, piezoelectric, mechanical, and other properties. Some physical properties of ceramics such as electrical and thermal conduction, superconductivity, and magnetic, dielectric, piezoelectric, and mechanical properties can be enhanced by controlled texture and grain orientation.[1-3] However, in ceramic processing, controlling of texture is more difficult than metals and polymers. In metals and polymers, texturing is achieved mainly by plastic deformation and slip plane rotation. However, in ceramics, which lack such plasticity, texture can be developed only by grain rotation or oriented, anisotropic grain growth.[4] More often, textured ceramics are achieved by orienting a second phase such as fibers, whiskers, and platelets during green body or powder consolidation processing.[5, 6]

In ceramics generally two principal techniques are known for texturing: 1) hot-working, and 2) Templated Grain Growth (TGG). Hot working or hot pressing is a cost effective method. The advantages of TGG technique are low cost and possibility of net-shape fabrication. In this method platelet, needle-like, or tabular seeds (single crystal particles) generally are produced by molten salt method and referred as isometric templates. The templates can be used in texturing of advanced ceramics such as structural ceramics and electroceramics.[1-3, 5]

On the other hand, The high piezoelectric coefficients available in <001> oriented single crystals in the relaxor ferroelectric -$PbTiO_3$ solid solution family have raised interest in polycrystalline ceramics and thin films of these compounds.[2, 7] It has been shown that extremely large piezoelectric coefficients (>2000 pC/N) can be achieved in rhombohedrally-distorted single crystals of this family.[7, 8] The piezoelectric properties of relaxor ferroelectric-$PbTiO_3$ depend on crystal orientations. For example, d_{33} values of about 2500, 590, 130 pC/N have been reported for single crystal of lead magnesium niobate-lead titanate (67PMN-33PT) in (001), (110), and (111) orientations, respectively (Table I). [9]

Table I. The piezoelectric properties of PMN-PT single crystal [9].

Orientation	ε (Room Temperature, @1 KHz)	tan δ (%)	k_t	d_{33} (pCN-1)
(110)	3700	0.8	0.51	590
(111)	910	1.1	0.42	130
(001)	5500	0.5	0.62	2400

Rapid development is being made in the fabrication of large single crystals, especially in the PMN-PT system with <001> cut. However, there are still considerable difficulties such as cost, low growth rate, segregation, and inhomogeneity across large crystals. [10] Consequently, developing textured bulk ceramics in a preferred orientation is an advantage for providing higher properties. Therefore, textured (1-x)($Mg_{1/3}Nb_{2/3}$) O_3-x$PbTiO_3$ (PMN-PT) ceramics obtained by the templated grain growth (TGG) process can posses a significant percentage of the piezoelectric properties of the Bridgman grown single crystals with the potential of lower cost, if there are not significant issues to achieve reliable piezoelectric results.

In structural ceramics like alumina or mullite, microstructral texturing results in preferred morphology of grains in one dimension that provides anisotropy in mechanical behavior. For example, in longitudinal direction of textured grains, higher strength can be achieved. Moreover, changing the grain morphology and grain size in structural ceramics generally may result a change in crack propagation resistance or fracture behavior. However, in piezoelectric ceramics, achieving higher microstructural texture may not necessarily result in higher piezoelectric properties. There are some other factors that can affect the properties. For example, long sintering cycle for achieving higher texture in TGG may result in pore coarsening. Chemical homogenities is also important and use of templates of the different phase with the PMN-PT matrix such as $SrTiO_3$ or $BaTiO_3$ may result in degraded piezoelectric properties.

The aim of this study is to address use of TGG for texturing PMN-PT ceramics. The issues in synthesizing different templates are discussed and the templates are compared regarding to their morphologies, sizes, and orientations. The viable template candidates for texturing of PMN-PT matrix are also demonstrated. Finally, the issues and perspective of the TGG technique for texturing of PMN-PT are presented.

EXPERIMENTS AND RESULTS

I) TEMPLATE SELECTION

Templates (seeds) act as nucleation sites for epitaxial growth of the matrix phase. There are two main criteria for a viable template: 1) crystallographic match and close unit cell parameters of templates and matrix, and 2) Template stability in TGG temperature.

According to first criteria, template particles must possess a similar crystal structure and <15% lattice parameter mismatch with the desired phase to be templated. [10] The PMN-PT as a matrix for TGG has a Perovskite structure with a unit cell parameter of $a_0 \approx 4.02$ Å (room temperature). Therefore, a template with perovskite structure and $a_0 = 4.02 \pm 0.6$ Å can be considered as viable templates. Here, the templates referred as single crystal seeds or particles, which can be synthesized by molten salt or other techniques. Although the PMN-PT single crystals can be fabricated, the synthesizing of PMN-PT seeds by molten salt, technically is an issue, which can be related to its phase diagram. Hence, progress is being made in seed synthesis of three types of titanates by molten salt or hydrothermal methods: [11-19] $SrTiO_3$ (a_0=3.90 Å), $BaTiO_3$ (a_0=3.99 Å), and $PbTiO_3$ (a_0=3.90 Å)*. The successful templated grain growth results in PMN-PT were reported using two templates: $SrTiO_3$ and $BaTiO_3$. Although $PbTiO_3$ is a viable template, hydrothermal synthesis of $PbTiO_3$ may not provide stable templates in TGG temperature and dissolution might happen. It should be noted, the successful results in TGG of perovskite PMN-PT were reported using templates synthesized by molten salt. [3, 16, 20-26] There is no evidence for TGG of perovskite structure by hydrothermally-synthesized seeds. This is due to instability of seeds at high temperatures during TGG and difficulties in obtaining a single crystal seeds from a low temperature process such as hydrothermal technique.

$SrTiO_3$ Templates

Most studies in grain orientation of PMN-PT matrix were devoted to strontium titanate (ST) templates. There are some reasons for selection of the template: 1) anisometric shape can be synthesized by two-step molten salt; 2) <001> texture in PMN-PT matrix can be achieved by <001> $SrTiO_3$ seeds.

In the TGG method, the initial orientation of the template in the matrix achieved by the shear stress results from the anisotropic shape during tape casting, which results in oriented grain growth. Thus, having a compatible anisotropically shaped template is of importance in the TGG method for successfully texturing ceramics. The template particles must have suitably high aspect ratio morphology (like a whisker, blade, or platelet), so that it can be mechanically oriented under an applied shear force during green forming. Although shape anisotropy is a useful characteristic for template orientation during processing, it is not necessary if other methods for alignment (e.g., a magnetic field) exist. [10]

Crystallographic orientations of seed surfaces are also important. In instances where the purpose of texturing is to access physical properties that are directly related to crystallographic orientation (like thermal conductivity, dielectric permittivity, piezoelectricity, electrical conductivity, etc.), it is preferable if the template axis matches the desired crystallographic orientation. As demonstrated in Table I, the enhanced piezoelectric properties can be achieved in <001> orientation.

Molten salt synthesis has produced cubic, tabular, and platelet morphologies of <001> perovskite $SrTiO_3$ seeds. [14, 17, 26] In our experiments, plate-like morphology with high aspect ratio of about 7-10 and an edge length of about 10-40 μm were achieved for <001> $SrTiO_3$ after epitaxial growth in a second molten KCl bath (Fig. 1a). Tabular morphology is also synthesized (Fig. 1b).

The reason of achieving different morphologies in the same molten salt cycles can be related to some factors such as purity and crystallographic phases of starting powders, dopants, and type and ratio of salts. Molten salt parameters like cooling and heating rates, soaking temperature, and time also play roles in size and shape of seeds.

To examine the crystallographic orientation of the seeds by XRD, two samples were prepared by tape casting of the slurries containing platelet $Sr_3Ti_2O_7$ and $SrTiO_3$ particles. According to the XRD patterns (Fig. 2a and 2b), two crystallographic planes of (110) and (200) in both layered perovskite $Sr_3Ti_2O_7$ and perovskite $SrTiO_3$ have the topotactic relations. Hence, there is a possibility of epitaxial growth of $SrTiO_3$ on these planes of $Sr_3Ti_2O_7$ core particles. Fig. 3b implies that the epitaxial growth was well developed on (200) plane, which corresponds to

<001> direction in $SrTiO_3$. If (110) pick is enhanced for epitaxial $SrTiO_3$ seeds which is synthesized in second step of molten salt, tabular or cubic morphologies with (110) direction (for cubic) and (110) and (200) directions (for tabular) perpendicular to the plate face of seeds might be expected. On the other hand, by strengthening of (200) direction for epitaxial $SrTiO_3$ seeds, platelet morphology with <001> orientation perpendicular to the plate face of seeds can be achieved.

Figure 1. SEM micrographs of the $SrTiO_3$ synthesized from $Sr_3Ti_2O_7$ and TiO_2 in molten KCl at 1100°C for 2 h. (a) platelet (rutile batch), and (b) tabular (anatase batch) morphologies.

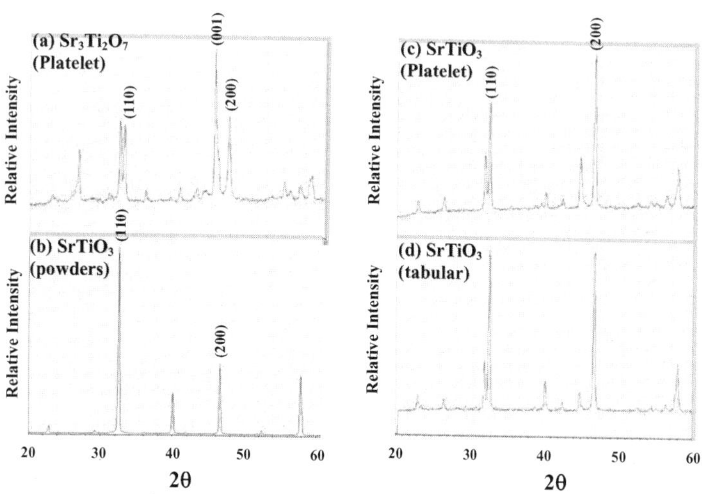

Figure 2. XRD patterns of (a) $Sr_3Ti_2O_7$ seeds, (b) $SrTiO_3$ powders (random orientations), (c) platelet $SrTiO_3$, and (d) tabular $SrTiO_3$.

$BaTiO_3$ Templates

$BaTiO_3$ seeds are good templates for the texturing of PMN-PT matrix because of similar crystallographic structure and unit cell parameter. However, there are some issues in using the seeds. According to literature [18, 22-24, 27] and our experiments, three types of templates were used for texturing of PMN-PT matrix: 1) <001> single crystal pieces synthesized by Remeika process, 2) <111> BaTiO3 seeds, and 3) <001> BaTiO3 seeds.

Here, the related issues and results for each template are explained.

i) <001> BT by Remieka process:

In our experiments <001> $BaTiO_3$ single crystal pieces were synthesized by Remeika process [28] (Fig. 3). The term of "seeds" may not be applied for these single crystals because of wide range of crystal sizes and high edge lengths of about 3-10 mm (Fig. 3a), which are too coarse for uniform distribution in matrix. Therefore, the pieces with those sizes should be crushed to proper sizes before incorporating in PMN-PT matrix.

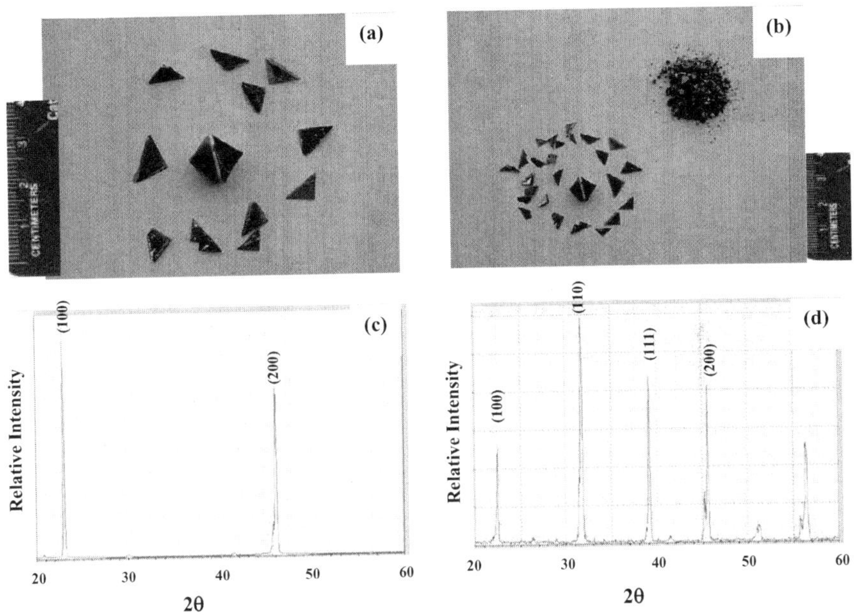

Figure 3. Photographs and XRD patterns of BT single crystal substrates and particles synthesized by Remieka process; (a) large BT substrates in triangle shape, (b) large BT substrates (bottom) and small BT particles (top); (c) XRD of BT substrates, (d) XRD of BT particles.

The single crystals, which were resulted in the process, can be divided to "substrates" (large triangle shape single crystals in Fig. 3a and in bottom of Fig. 3b) and "particles" (small single crystals in top of Fig. 3b). The <001> substrates can be used for Seeded Polycrystal Conversion (SPC) substrates [3] or may be crushed for using as templates for TGG [26]. The BT substrates have excellent picks of <001> orientation (Fig. 3c) without significant pick of (110), which is generally the strongest pick in $BaTiO_3$ powder (random orientation). Fig. 3d indicates the XRD pattern from $BaTiO_3$ particles (top of Fig. 3b); Although the (100) and (200) picks are stronger than those of BT random orientation, (110) and (111) picks are strong and appreciable.

Therefore, the particles may not be well aligned in PMN-PT matrix because of lower anisometric shape and also may result lower <001> texturing compare with crushed substrates as templates. Moreover, the results of using BT substrates or crushed BT as templates in PMN-PT matrix revealed that the growth rate on <001> surface of the BT templates is not as fast as <001> $SrTiO_3$ seeds (in the same TGG temperature).

ii) <111> BT Seeds:

These types of seeds can be synthesizes by a two-step molten salt. In our experiments, platelet $Ba_6Ti_{17}O_{40}$ seeds were synthesized in first step. High aspect ratio (>15) with a diameter of 20-50 μm was achieved (Fig. 4a). There is a topotactic relation between $Ba_6Ti_{17}O_{40}$ and $BaTiO_3$: $(008)_{B6T17}//(111)_{BaTiO3}$. Therefore, platelike BaTiO3 particles with (111) direction perpendicular to the plate face can be obtained. It means that the plate faces of $BaTiO_3$ are (111). As mentioned before, although <111> direction provides higher growth rate because of higher atomic density compared with <001> direction, this direction cannot provide higher piezoelectric properties in PMN-PT matrix compared with <001> direction (Table I). Therefore, in this respect, texturing of PMN-PT matrix in <111> orientation is not preferred.

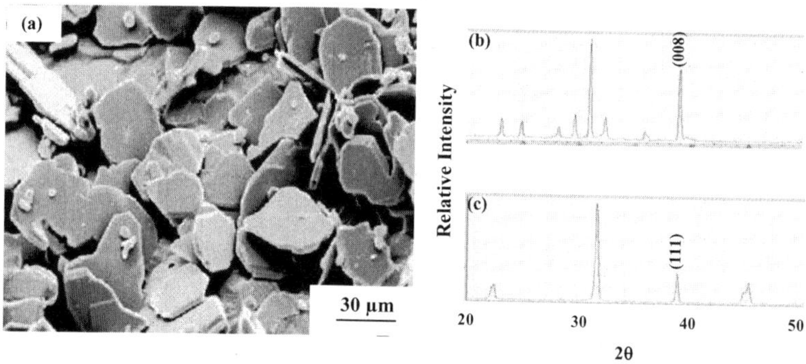

Figure 4. (a) SEM micrograph and (b) XRD pattern of $Ba_6Ti_{17}O_{40}$ seeds. (c) XRD pattern of $BaTiO_3$ powder (random orientation).

iii) <001> BT Seeds:

Liu et al. [19] recently used a two-step molten salt process to synthesize <001> BT seeds. Richter et al. [23] also used this technique to synthesize BT seeds and they used the seeds to texture PMN-PT and PMN-PZT matrixes by TGG. They reported high texture with Lotgering factors between 0.94-0.99 and enhanced dielectric and piezoelectric properties (0.25% strain at 3 kV/mm and large d_{33} of up to 878 pm/V). In both studies, <001> BT seeds were synthesized according to following reactions:

$$2Bi_2O_3 + 3TiO_2 \rightarrow Bi4Ti_3O_{12} \quad (1)$$

$$Bi4Ti_3O_{12} + BaCO_3 + 4TiO_2 \rightarrow BaBi_4Ti_4O_{15} + CO_2 \quad (2)$$

$$BaBi_4Ti_4O_{15} + BaCO_3 \rightarrow 2BaTiO_3 + 2Bi_2O_3 + CO_2 \quad (3)$$

The reactions (1) and (2) were used in first step and reaction (3) was used in second step to synthesized epitaxial BT seeds by converting platelet $BaBi_4Ti_4O_{15}$ seeds to platelet $BaTiO_3$ seeds. Fig. 5 shows the XRD patterns of $BaBi_4Ti_4O_{15}$ and $BaTiO_3$ compounds. According the XRD pattern, There is a topotactic relation between $BaBi_4Ti_4O_{15}$ and $BaTiO_3$: $(0018)_{BaBi4Ti4O15}//(111)_{BaTiO3}$. Therefore, again platelike $BaTiO_3$ seeds with (111) direction perpendicular to the plate face may be expected. It means that the plate faces of $BaTiO_3$ may not be <001>.

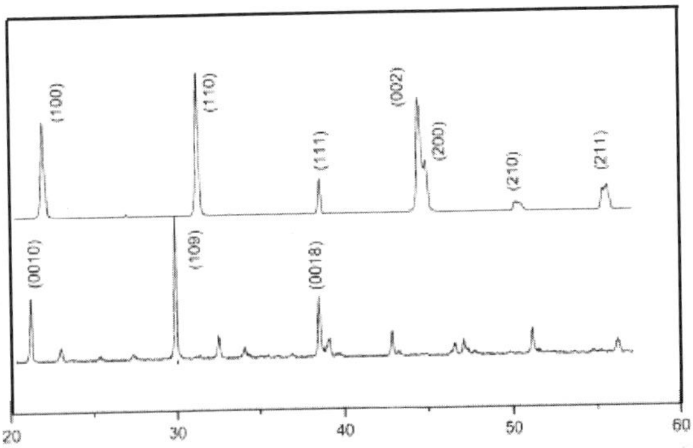

Figure 5. XRD patterns of $Ba_6Ti_{17}O_{40}$ seeds. (c) XRD pattern of $BaTiO_3$ powder (random orientation).[19]

Other Templates

Some other viable templates with perovskite structure and close unit cell parameter to PMN-PT may be synthesized to texture PMN-PT matrix. As it mentioned before the templates with a unit cell parameter in the range of a_0=4.02±0.6 Å can be considered as viable templates for texturing of PMN-PT matrix. Calcium Titanate (CT) has a perovskite structure with a_0= 3.795 Å in room temperature. Therefore, this system was also studied. Fig. 6 illustrates the phase diagram of $CaTiO_3$. The existence of two stoichimetry compounds (other than $CaTiO_3$), between TiO_2 and CaO, provides the possibility of synthesizing of $Ca_3Ti_2O_7$ and $Ca_4Ti_3O_{10}$ seeds by molten salt as cores for epitaxial CT. In our experiments, $Ca_3Ti_2O_7$ seeds were synthezied by molten salt as cores for epitaxial CT (Fig. 7a). Some platelet $Ca_3Ti_2O_7$ seeds with edge length of ~40 μm were resulted. However, the performance of process is low and needs to further work. Sait et al.[29] recently synthesized the platelet CT seeds by a topochemical microcrystal conversion method and the CT templates were used to fabricate textured microwave dielectric ceramics. They synthesized Platelike $CaTiO_3$ seeds by a two-step molten salt process (Fig. 7b). In the first step, $CaBi_4Ti_4O_{15}$ platelet seeds were synthesized and were converted to Platelet $CaTiO_3$ in the second step. When the templates used in $CaTiO_3$ matrix, high <001> orientation (99.3%) was reported.[29]

There is not evidence up to now to show using the CT templates for texturing of PMN-PT matrix. However, its close unit cell parameter to PMN-PT and results of <001> orientation in a perovskite matrix (like CT), the CT templates may be considered as promising templates for TGG of PMN-PT ceramics.

In the case of PMN-PT, anisometric seeds of this relaxor ferroelectric phase have not been obtained up to now, and textured PMN-PT ceramics have been prepared using alternative anisometric templates such as $SrTiO_3$ and $BaTiO_3$, which is normally called heterogeneous-TGG. Although, the use of templates of the same phase is preferred, there are technical issues in synthesizing of anisometric PMN-PT seeds.

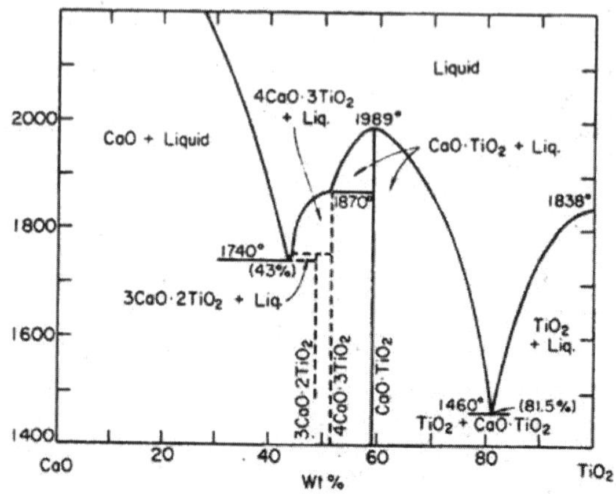

Figure 6. Phase diagram of $CaTiO_3$ system. [30]

Figure 7. (a) $Ca_3Ti_2O_7$ seeds as cores for epitaxial CT; (b) The platelet CT seeds synthsized by Sait et al. [29].

II) TEMPLATED GRAIN GROWTH

The results show microstructural texturing and grain orientation in PMN-PT matrix by using ST and BT templates. High <001> grain orientations (>90%) were reported by using

<001> templates [20-27]. For example, in our experiments, high Lotgering factors (>70%) were obtained when <001> platelate ST templates were used to texture PMN-PT matrix by tape casting or Fused Deposition of Ceramics (FDC) (Fig. 8). However, some issues such as properties degradation and porosity development due to long thermal TGG process, difficulties in achieving high density in result of void formation and difference in template shape and size with matrix, chemical composition instability in long time sintering, interaction of ST seeds and PMN-PT matrix, different ferroelectric properties of templates and matrix, and high fluctuations in piezoelectric properties, may face templated grain growth (TGG) with several problems as a promising technique for texturing of PMN-PT matrix. Although the high amount of texture may mean good stability and performance of templates, it may not provide improved piezoelectric properties and even can result degraded properties.

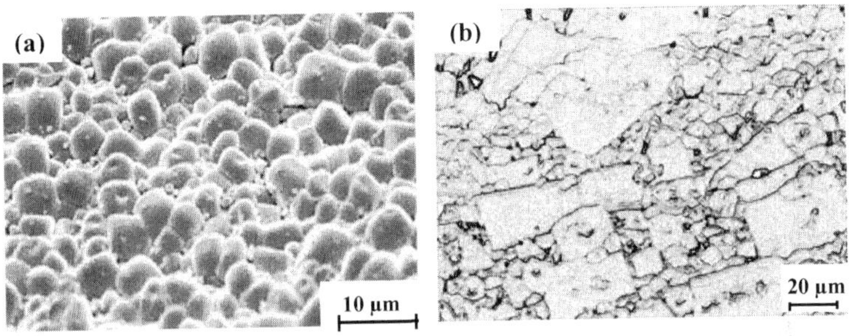

Figure 8. a) Fine equiaxed microstructure of unseeded, and b) Grain orientation of seeded PMN-35PT matrix sintered at 1150 °C for 5h.

Therefore, regarding to microstructural texturing of PMN-PT matrix, the results of TGG by <001> ST templates, and <001> BT crushed substrates, and <111> BT templates show high grain orientation (70-90%) [21-23, 26, 27]. It is noteworthy that up to now only <001> ST templates have technical importance for TGG of PMN-PT, because of potentially high piezoelectric properties in <001> orientation, and relative ease of processing of anisometric <001> orientation. Nevertheless, incorporating the ST templates in PMN-PT matrix may not provide reliable piezoelectric properties. There are some issues that should be considered. Longer firing time is needed for TGG samples compared with conventional random orientation PMN-PT ceramics. The sintering times of 5-10 h were used to achieve high texturing which can result some problems in this relaxor ferroelectric phase such as variable compositions, pore coarsening, grain impingement, grain boundary phases, difficulties in obtaining full densities, and dissolving and interaction of ST templates with matrix. There is a gradient in the size of the pores in the grown crystals which is a result of the coarsening of the porosity in the matrix during heating.[10] It should be noted the <001> orientation has slowest growth rate compared with <110> and <111> orientations in TGG which needs longer time of heating to achieve high texture. On the other hand, template composition and induced changes in the domain stability (especially for compositions near a morphotropic phase boundary) can play an important role in reliable properties. Consequently, above mentioned issues and other problems might lead to degradation of piezoelectric properties.

CONCLUSION

In this paper, synthesizing of different viable templates for texturing the PMN-PT ceramics including strontium titanate (ST), barium titanate (BT), and calcium titanate (CT) were discussed. The anisometric template morphologies and their crystallographic orientations were compared and viability the templates for microstructural texturing of PMN-PT matrix was studied. ST and BT were successfully used to texture PMN-PT matrix. However, regarding to improved piezoelectric properties, there are some issues such as microstructural degradation and interaction between seeds and matrix, which might lead to degradation and high fluctuations in piezoelectric properties in templated grain growth. Although, the templated grain growth (TGG) can be considered as an effective technique in low cost for improvement of properties of structural ceramics and some ferroelectric and piezoelectric ceramics, there are some issues for using this technique for PMN-PT ceramics.

FOOTNOTES

* Unit cell parameters in room temperature.

REFERENCES

[1]M. M. Seabaugh, I. H. Kerscht, and G. L. Messing, Texture Development by Templated Grain Growth in Liquid-phase-sintered α-Alumina, *J. Am. Ceram. Soc.,* **80**, 1181–1188 (1997).

[2]S. Trolier-McKinstry, E. Sabolsky, S. Kwon, C. Duran, T. Yoshimura, J.-H. Park, Z. Zhang, and G.L. Messing, in *Piezoelectric Materials in Devices*, edited by N. Setter, 1st ed. (EPFL, Lausanne, Switzerland), p. 497-518 (2002).

[3]M. Allahverdi, A. Hall, R. Brennan, M. E. Ebrahimi, N. M. Hagh, and A. Safari, An Overview of Rapidly Prototyped Piezoelectric Actuators and Grain-Oriented Ceramics, *Journal of Electroceramics* **8** 129–137 (2002).

[4]M. S. Sandlin and K. J. Bowman, Textures in AlN–SiC Composite Ceramics, in Materials Research Society Symposium Proceedings, Vol. 327, *Covalent Ceramics II, Non-Oxides.* Material Research Society, Pittsburgh, PA, p. 263–68 (1994).

[5]S. H. Hong and G. L. Messing, Development of Textured Mullite by Templated Grain Growth, *J. Am. Ceram. Soc.,* **82**, 867–872 (1999).

[6]P. F. Becher, Microstructural Design of Toughened Ceramics," *J. Am. Ceram. Soc.*, **74,** 255–69 (1991).

[7]S-E. Park and T.R. Shrout, Ultrahigh Strain and Piezoelectric Behavior in Relaxor Based Ferroelectric Single Crystals, *J. Appl. Phys.*, **82,** 1804 (1997).

[8]J. Kuwata, K. Uchino, and S. Nomura, *Jpn. J. Appl. Phys.,* **21,** 1298 (1982).

[9]G. Xu, H. Luo, Y. Guo, Y. Gao, H. Xu, Z. Qi, W. Zhong, Z. Yin, Growth and piezoelectric Properties of $Pb(Mg_{1/3}Nb_{2/3})O_3$-$PbTiO_3$ Crystals by Modified Bridgman Technique, *Solid State Communications,* **120,** 321 (2001).

[10]G.L. Messing et al., Templated Grain Growth of Texured Piezoelectric Ceramics, *Critical Reviews in Solid State and Materials Sciences*, **29,** 45-96 (2004).

[11]H. Cheng, J. Ma, Z. Zhao, and D. Qiang, *J. Am. Ceram. Soc.*, **75,** 1123 (1992).

[12]J. Moon, J.A. Kerchner, J. LeBleu, A.A. Morrone, and J.H. Adair, *J. Am. Ceram. Soc.*, **80,** 2613 (1997).

[13]T. Takeuchi, T. Tani, and T. Satoh, *Solid State Ionics*, **108,** 67 (1998).

[14]K. Watari, B. Brahmaroutu, G.L. Messing, and S. Trolier-McKinstry, *J. of Mat. Res.*, **15,** 846 (2000).

[15]Y. Ito, B. Jadidian, M. Allahverdi, and A. Safari, *Proceedings of the 12th IEEE Int. Symp. on Application of Ferroelectrics*, edited by S.K. Streiffer, B.J. Gibbons, and T. Tsurumi (The institute of Electrical and Electronic Engineers, Piscataway, NJ), Vol. 1, p. 389 (2000).
[16]E.M. Sabolsky, A.R. James, S. Kwon, S. Trolier-McKinstry, and G.L. Messing, *J. Applied Phys. Letters,* **78, 2551** (2001).
[17]M.E. Ebrahimi, M. Allahverdi, A. Safari, Synthesis of High Aspect Ratio Platelet $SrTiO_3$, *J. Am. Ceram. Soc.,* **88,** 2129 (2005).
[18]M.E. Ebrahimi, M. Allahverdi, A. Safari, High Aspect Ratio Platelet $SrTiO_3$ for Templated Grain Growth of PMN-PT Ceramics, *Ceram. Transactions,* **136,** 241-250 (2003).
[19]D. Liu, Y. Yan, and H Zhou, Synthesis of Micron-Scale Platelet $BaTiO_3$, *J. Am. Ceram. Soc.*, **90,** 1323–1326 (2007).
[20]S. Kwon, E.M. Sabolsky, G.L. Messing, S. Trolier-McKinstry, *Proceedings of the 10th US-Japan Seminar on Dielectric and Piezoelectric Ceramics,* Providence, RI, Vol. 1, p. 327 (2001).
[21]M.E. Ebrahimi, M. Allahverdi, and A. Safari, Platelet $SrTiO_3$ seed synthesis for texturing of PMN-PT matrix, *at 2002 US Navy Workshop on Acoustic Transduction Materials and Devices*, Baltimore, Maryland, May 13-15, (2002).
[22]R. Brennan, M. Allahverdi, M. E. Ebrahimi, and A. Safari, Growth of net-shape PMN-PT single crystal components by Fused Deposition of Ceramics (FDC) and templated grain growth (TGG) techniques, *in Proceedings of the 13th IEEE International Symposium on Applications of Ferroelectronics (ISAF)*, Nara, Japan, 435-438 (2002).
[23]T. Richter, S. Denneler, C. Schuh, and E. Suvaci, Textured PMN–PT and PMN–PZT, *J. Am. Ceram. Soc.*, **91,** 929–933 (2008).
[24]T. Kimura, Y. Miura, K. Fuse, "Texture Development in Barium Titanate and PMN-PT using Hexabarium 17-titanate Hetrotemplates," *Int. J. Appl. Ceram. Technol.*, **2,** 15-23 (2005).
[25]E.R.M. Andreeta, H.F.L. dos Santo, M.R.B. Andreet, M.H. Lente, D. Garcia, A.C. Hernandes, and J.A. Eiras, Anisotropy on SrTiO3 templated textured PMN–PT monolithic ceramics, *Journal of the European Ceramic Society,* **27,** 2463–2469 (2007).
[26]S. Kwon, E. M. Sabolsky, G. L. Messing, and S. Trolier-McKinstry, High Strain, <001> Textured $0.675Pb(Mg_{1/3}Nb_{2/3})O_3$–$0.325PbTiO_3$ Ceramics: Templated Grain Growth and Piezoelectric Properties, *J. Am. Ceram. Soc.*, **88,** 312–317 (2005).
[27]E.M. Sabolsky, S. Trolier-McKinstry, and G. L. Messing, Dielectric and piezoelectric properties of <001> fiber-textured 0.675Pb(Mg1/3Nb2/3)O3– 0.325PbTiO3 ceramics, *J. of Appl Phys.,* **93,** 4072-4080 (2003).
[28]J. P. Remeika, *J. Am. Chem. Soc.*, **76**, 940 (1954).
[29]Y. Sait, H. Takao, and K. Wada, Synthesis of Platelike $CaTiO_3$ particles by a Topochemical Microcrystal Conversion Method and Fabrication of Textured Microwave Dielectric Ceramics, *Ceramics International*, **34,** 745–751 (2008).
[30]E.M. Levin, C.R. Robbins, and H.F. McMurdie, Phase Diagrams for Ceramists, Published by *Am. Ceram. Soc.*, Ohio, Columbus (1989).

ELECTRICAL CHARACTERIZATION AND DIELECTRIC RELAXATION OF AU/POROUS SILICON CONTACTS

M. Chavarria and F. Fonthal

Advanced Materials for Micro and Nanotechnology Research Group
Facultad de Ingeniería, Universidad Autónoma de Occidente
Km. 2 vía Cali - Jamundi, Valle del Lili, Cali, Colombia

ABSTRACT

The DC and AC electrical characterization of Au/porous silicon contacts in room temperature is presented. The Porous Silicon layers were prepared by electrochemical etching in p-type silicon <100> substrates. The DC resistances were studied and the AC electrical measurements were performed from 5 Hz to 10 MHz, for the four samples at 0V. We found two behaviours typical of the dielectric permittivity property of samples studied; i) the regimen of strong dispersion present at low frequency and ii) the relaxation region that dominates at high frequency. An electrical equivalent circuit was proposed to fit the experimental frequency response of the different samples. We have obtained various model parameter values fitting corresponding Au/PS structures fabricated under different process conditions.

INTRODUCTION

Since the first work by Koshida et al. [1] there have been published several papers dedicated to the DC and AC electrical characterization of porous silicon layers [2,3]. Porous silicon (PS) is a material that has attracted a great deal of attention because of its low dimensional semiconductor structure and its potential applications, such as photoluminescence [4], temperature sensor [5], gas sensor [6,7], and photodetectors [8,9]. Nowadays, the electrical impedance methods are to be employed on a larger scale in the fabrication of the electronic components on the basis of micro-machined silicon and porous silicon.

Measurement of DC and AC electrical characterization is important for dielectric characterization of materials. This characterization technique permits to separate the dielectric permittivity properties corresponding to the capacitances present in the relaxation region in the porous silicon layers [10]. The AC dielectric analysis is interpreted in terms of the admittance or impedance measurement and the equivalent electrical circuits are formed by RC networks in parallel, connected in series [11,12]. In conjunction with structural characterization, the various model parameter values fitting corresponding Au/PS structures fabricated completed a physical analysis in the structure studied under different fabrication processes [13,14,15].

An important aspect that should be addressed to enhance the electrical performance for these devices is the analysis of the electrical contacts on the PS layer. Various authors have reported different conduction mechanisms involved in metal-PS devices, depending on the fabrication process. Among them, Bohn et al [9] reported that the transport mechanism is controlled by two Schottky junctions symmetric in both voltages for metal-semiconductor-metal photoconductor, Theodoropoulou et al [15] presented that the Ohmic conduction is dominated by the bulk resistance, Balagurov et al [16,17] reported the power law Space charge limited current (SCLC) in the conduction mechanism and Ben-Chorin et al [18,19] have shown that DC conductivity is due to the Poole–Frenkel mechanism.

In this paper, we present the AC impedance analysis at room temperature and the DC current-voltage characteristics of an Au/PS structure. Finally, we have found an equivalent circuit which properly fits the AC experimental measures of the structure studied.

EXPERIMENTS AND CHARACTERIZATION

P–type silicon wafers with <100> orientation and resistivities of 4-7 Ω·cm and 7-9 Ω·cm were used as starting material. An Aluminum contact was sputter-deposited on the backside of the wafer.

The electrolyte was prepared to the desired concentration by adding 50 % HF to ethanol (1:1) solvent mixture. The wafer was mounted in an electrochemical cell with the front side in contact with the electrolyte. The current density was 5 mA/cm^2 and the etching times were 90 s and 180 s (see the samples in the Table 1). Finally, thin Au spots were evaporated on the top of the porous silicon layer in order to obtain the electrical contacts (The diameter of the spots was 0.75 mm and the separation between them was 2.5 mm). Figure 1 shows the schematic view and the SEM micrographs of the Au/PS/Au structures developed in this work.

An HP 4145B parameter analyzer was used to measure the DC current – voltage characteristic. The AC electrical conductivity was measured with an HP4192A Impedance Analyzer between 5 Hz and 10 MHz and the zero voltage was measured at room temperature. Figures 2 show the experimental impedance values in modulus (Figure 2a) and phase φ (Figure 2b) as functions of frequency for the Au/PS/Au structures. We found that all samples presented the DC behaviour characteristic of semiconductors at low frequency and in the sample C, was presented a phase of -85° at 20 KHz compared with the others samples that presented a phase of -90° at 500 KHz. This difference is interesting because with the same etching time and different resistivity is possible to dielectric permittivity change in porous silicon.

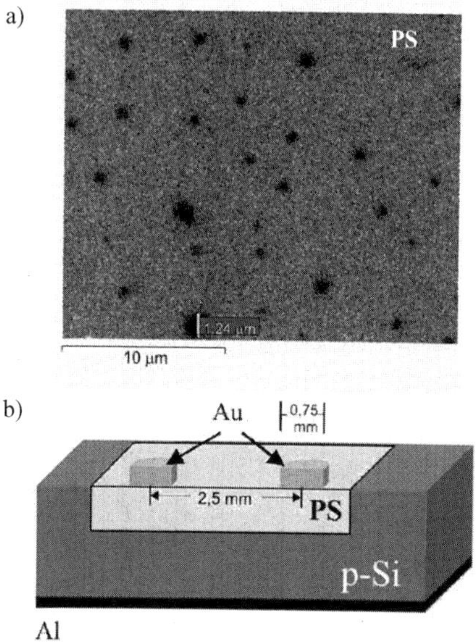

Figure 1. a) SEM micrographs of the samples fabricated and b) Schematic structure of the Au/PS contacts.

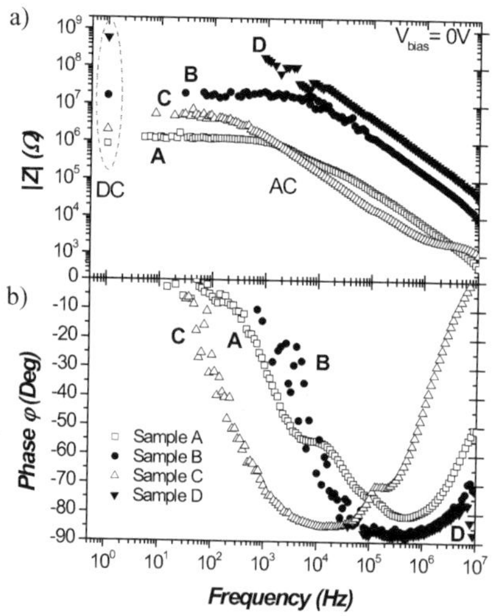

Figure 2. Z_{eq} modulus a) and phase φ (deg) b) plots the experimental curve.

RESULTS AND DISCUSSION

In the Figure 3, shown the electrical AC characteristic of the porous silicon thin films with the conductance and the capacitance properties [13]. We found at low frequency the barrier conductivity in the conductance corresponding to the DC value determined by the I-V characteristic and at high frequency we obtained the purely conductance value of the PS layers, can been see in the Figure 3a the two behaviours.

The capacitance observed in the Figure 3b corresponds to the barrier capacitance in the diffusion region (dispersion) at low frequency and in the relaxation region corresponds to the geometry capacitance (C_0) at high frequency.

The dispersion presented in the capacitance measurement at low frequency in Figure 3b, is usually attributed to the Maxwell – Wagner phenomenon due to the contact polarization (parasitic capacitance) [12]. The capacitance values decrease for the samples with lower etching time and the conductance values increase for the sample with higher etching time or lower resistivity. The frequency dependence of both measurements is quite different. In the Figure 3 the fitting performed (lines) were shown, we modelled the equivalent circuit shown in the Figure 4 by equation (1) for the samples studied.

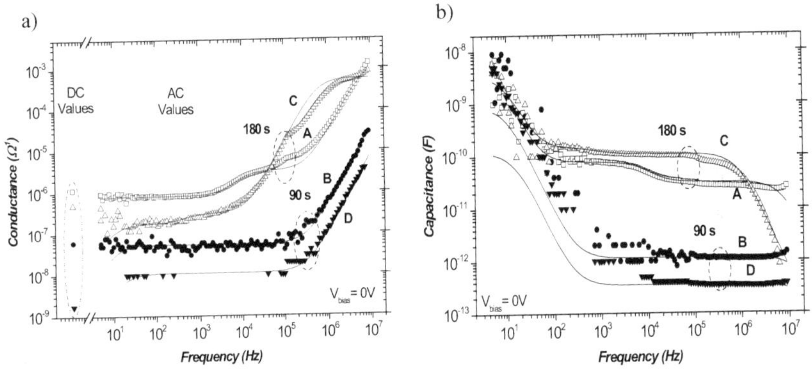

Figure 3. a) Conductance and b) capacitance vs. frequency dependencies of Au/PS/Au structures studied for different resistivities and etching times at room temperature (Symbols are data experiment and lines are fitting values based on Eq (1)).

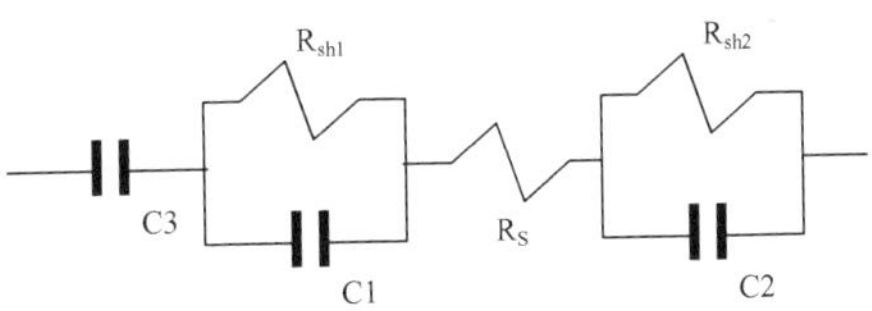

Figure 4. Shows the electrical equivalent circuit.

Figure 4 shows the electrical equivalent circuit used for the fitting of the experimental measurements. It is based in two back-to-back Schottky diodes and consists of a combination of two parallel RC

networks ($R_{sh}C$ network- due to the depletion capacitance and the shunt resistance) in series with a series RC network (Series resistor (R_s) an important parameter for high frequencies > 1 MHz and the Capacitance (C_3) for low frequencies).
Due to the dimensions of the studied devices, the geometrical capacitance of the porous silicon layer influence in the electrical behaviour the Au/PS/Au structures studied with higher etching time. The frequency dependence of the equivalent impedance Z_{eq} is:

$$Z_{eq}(\omega) = R_s + \frac{R_{sh1}}{(iR_{sh1}\omega C_1 + 1)} + \frac{R_{sh2}}{(iR_{sh2}\omega C_2 + 1)} + i\frac{1}{\omega C_3} \tag{1}$$

where ω is the angular frequency of AC signal. From this, determine the real part and the imaginary part of the complex impedance:

$$\mathrm{Re}_{(z)} = R_S + \frac{R_{sh1}}{C_1^2 R_{sh1}^2 \omega^2 + 1} + \frac{R_{sh2}}{C_2^2 R_{sh2}^2 \omega^2 + 1} \tag{2}$$

$$\mathrm{Im}_{(z)} = \frac{1}{\omega C_3} + \frac{C_1 R_{sh1}^2 \omega}{C_1^2 R_{sh1}^2 \omega^2 + 1} + \frac{C_2 R_{sh2}^2 \omega}{C_2^2 R_{sh2}^2 \omega^2 + 1} \tag{3}$$

We are fitting the samples with high etching time using the complete equivalent circuit (two parallel RC networks and a series RC network), but for the fitting, the samples with low etching time are using only one parallel RC network. We obtained that the shunt resistance increases when the resistivity increases and the etching time decreases, so we found higher values of shunt resistance for the samples with lower etching time (one parallel RC network). Also the capacitances increase if the resistivity and the etching time increase. The table 1 shows the parameters fitting that used the equivalent circuit of the Figure 4.

Table 1. Fitting parameters according to Eqs. (2), (3) and (6) for different resistivities and etching times at 0V.

Sample	Etching Time (s)	ρ (Ωcm)	R_{sh1} (kΩ)	C_1 (pF)	R_{sh2} (kΩ)	C_2 (pF)	C_3 (nF)	R_S (kΩ)	B'	α
A	180	4 - 7	146	98,00	1060	45	100	0,7	24869,3	0,96
B	90	4 – 7	18000	1,15	-	-	0,9	9,0	37472,0	0,98
C	180	7 – 9	500	450,00	4500	250	3,5	1,8	633,7	0,85
D	90	7 - 9	89900	0,35	-	-	0,1	20	101733,3	1,01

The complex admittance $Y_{eq} = 1/Z_{eq}$ can be converted into the complex permittivity formalism ε_{eq} by the relation [11]:

$$\varepsilon_{eq} = \varepsilon' - j\varepsilon'' = \frac{Y_{eq}}{j\omega C_0} \tag{4}$$

$$\varepsilon' = \frac{\mathrm{Im}_{(Z)}|Y|^2}{\omega C_0}, \quad \varepsilon'' = \frac{\mathrm{Re}_{(Z)}|Y|^2}{\omega C_0}, \quad C_0 = \varepsilon_0 \frac{A}{l} \tag{5}$$

where ε' and ε'' represent the real and the imaginary parts of the permittivity of vacuum, respectively. |Y| is the admittance modulus and C_0 is the geometric capacitance present at high frequency.

Based on the universal dynamic law proposed by Dr. Jonscher [20], the imaginary part of the dielectric property can be expressed as [11,18]:

$$\varepsilon''(\omega) = B'(T)\omega^{-\alpha} \qquad (6)$$

where B' is a constant and α is an exponent that determines the frequency dependence of the dielectric.

The frequency dependence of the real part and the imaginary part of the dielectric constant is shown in the Figure 5. Contact polarization effect can be seen in the thickness dependence of the dielectric constant. We found two behaviours present in the real part of the dielectric constant (Figure 5a). The Figure 5b shows the imaginary part of the dielectric property that decreases when the frequency increases linearly, presented the $\alpha \approx 1$ due to the frequency dependence [11,18] and we found the B values for the four samples at low frequency. At high frequency all samples presented the typical behaviour of the relaxation region ($\nu > 2$ kHz) [12], but the dielectric property changed when the frequency increased due to the low capacitance in the sample C and was found the relation between the dielectric constant and the thickness of the samples studied, due to the dielectric constant is higher for the samples that were fabricated with more etching time compare with the other samples.

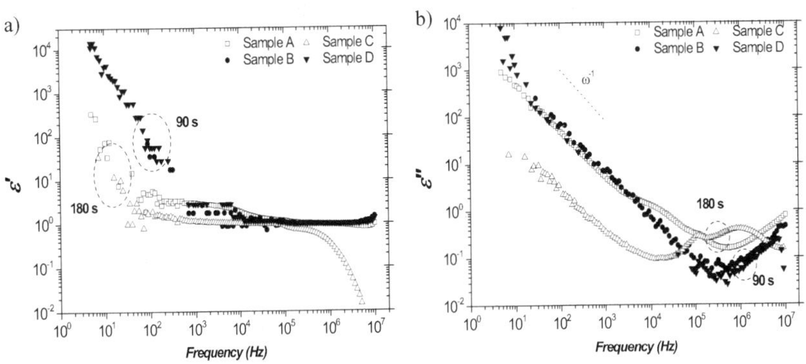

Figure 5. Frequency dependence of the real part a) and the imaginary part b) by the complex dielectric function at room temperature.

CONCLUSIONS

The electrical DC and AC characteristics of Au/porous silicon contacts were studied. An electrical equivalent circuit which describes the AC impedance measurements has been proposed. We are fitting the samples with high etching time (180 s) using the complete equivalent circuit, but for the fitting of the samples with low etching time (90 s) we used only one parallel RC network.

We have found the relationship between the shunt resistance (R_{sh}) and capacitance (C) to the resistivity (ρ) or etching time (Et). The R_{sh} increased for the samples with higher ρ and lowered Et. However C increased if the ρ or the Et increased.

We have determined the α exponent (close to 1) that explains the frequency dependence of the imaginary part the dielectric property. We obtained the typical characteristic of the basic model circuit with one RC parallel network and the series resistance for the samples with lower etching time.

ACKNOWLEDGEMENTS

The authors would like to thank Dr. A. Rodriguez (Universitat Politècnica Catalunya) and Dr. J. Pallarès (Universitat Rovira i Virgili) for their support in this investigation. This work was supported by the Spanish Commission of Science and Technology (CICYT) under Grant No. TIC2005-02038 and by the Universidad Autónoma de Occidente (UAO) under Project No. 07INTER-79.

REFERENCES

[1] N. Koshida, M. Nagasu, K. Echizenia, and Y. Kiuchi, Impedance Spectra of p-Type Porous Si-Electrolyte Interfaces, *J. Electrochem. Soc.*, **133**, 2283 – 2287 (1986).

[2] M. Ben-Chorin, F. Möller, and F. Koch, AC conductivity in porous silicon, *J. Luminescence*, **57**, 159 – 162 (1993).

[3] V. Parkhutik, E.S. Matveeva, F. Namavar, N. Kalcoran, Mechanism of AC Electrical Transport of Carriers in Freshly Formed and Aged Porous Silicon, *J: Electrochem. Soc.*, **143**, 3943 – 3949 (1996).

[4] L.T. Canham, Silicon quantum wire array fabrication by electrochemical and chemical dissolution of wafers, *Appl. Phys. Lett.*, **57**, 1046 – 1048 (1990).

[5] J. Salonen, M. Björkqvist, J. Paski, Temperature-dependent electrical conductivity in thermally carbonized porous silicon, *Sensors and Actuators A*, **116**, 438 – 441 (2004).

[6] S. Khoshnevis, R.S. Dariani, M.E. Azim-Araghi, Z. Bayindir, K. Robbie, Observation of oxygen gas effect on porous silicon-based sensors, *Thin Solid Films*, **515**, 2650 – 2654 (2006).

[7] F. Fonthal, T. Trifonov, A. Rodríguez, C. Goyes, X. Vilanova, J. Pallarès, Fabrication and characterization of porous silicon on crystalline silicon based devices, *Proceeding of the IEEE Computer Society, 4th Int. Conf. CERMA 2007*, **1**, 170 – 174 (2007).

[8] F. Fonthal et al, Electrical and optical characterization of porous silicon/p-crystalline silicon heterojunction diodes, *AIP Conference Proceeding, VI Iberoamerican. Conf RIAO 2007 and IX Lati american meeting OPTILAS 2007*, **992**, 780 – 785 (2008).

[9] A.M. Rossi, H.G. Bohn, Photodetectors from Porous Silicon, *Phys. Stat. Sol. (a)*, **202**, 1644 – 1647 (2005).

[10] E. Axelrod, A. Givant, J. Shappir, Y. Feldman, A. Sa'ar, Dielectric relaxation and transport in porous silicon, *Phys. Rev. B*,.**65**, 165429-1 – 7 (2002).

[11] L.K. Pan, H.T. Huang, Chang Q. Sun, Dielectric relaxation and transition of porous silicon, *J. Appl. Phys.*, **94**, 2695 – 2700 (2003).

[12] A.K. Jonscher, *Dielectric Relaxation in Solids*, Chelsea Dielectrics Press, London (1983).

[13] F. Fonthal, C. Goyes, A. Rodríguez, Electrical Transport and Impedance Analysis of Au/Porous Silicon Thin Films, *Proceeding of the IEEE Computer Society, 5th Int. Conf. CERMA 2008*, **1**, 3 – 7 (2008).

[14] F. Fonthal, T. Trifonov, A. Rodríguez, L.F. Marsal, J. Pallarès, AC impedance analysis of Au/porous silicon contacts, *Microelectronic Eng.*, **83**, 2381 – 2385 (2006).

[15] M. Theodoropoulou et al, Transient and ac electrical transport under forward and reverse bias conditions in aluminum/porous silicon/ p-cSi structures, *J. Appl. Phys.*, **96**, 7637 – 7642 (2004).

[16] L.A. Balagurov et al, Transport of carriers in metal/porous silicon/c-Si device structures based on oxidized porous silicon, *J. Appl. Phys.*, **90**, 4543 – 4548 (2001).

[17] L.A. Balagurov, S.C. Bayliss, A.F. Orlov, E.A Petrova, B. Unal, D.G. Yarkin, Electrical properties of metal/porous silicon/p-Si structures with thin porous silicon layer, *J. Appl. Phys.*, **90**, 4184 - 4190 (2001).

[18] M. Ben-Chorin, F. Möller, F. Koch, W. Schirmacher, M. Eberhard, Hopping transport on a fractal: AC conductivity of porous silicon, *Phys. Rev. B*, **51**, 2199 – 2213 (1995).
[19] M. Ben-Chorin, F. Möller, F. Koch, Nolinear electrical transport in porous silicon, *Phys. Rev. B.*, **49**, 2981 – 2984 (1994).
[20] A.K. Jonscher, *Universal Relaxation Law*, Chelsea Dielectrics Press, London (1996).

STRUCTURAL AND DIELECTRIC PROPERTIES OF THE $Na_{0.5}Bi_{0.5}TiO_3$–$NaTaO_3$ CERAMIC SYSTEM

Jakob König, Matja Spreitzer, Boštjan Jančar, and Danilo Suvorov
Advanced Materials Department, Jožef Stefan Institute,
Jamova 39, 1000 Ljubljana, Slovenia

ABSTRACT

X-ray diffraction analysis and scanning electron microscopy have revealed that the solid solutions between $Na_{0.5}Bi_{0.5}TiO_3$ and $NaTaO_3$ form across the whole concentration range. Additionally, morphotropic phase compositions were found for samples with 5 and 15 mol% of $NaTaO_3$. A study of the dielectric properties showed that with increasing content of $NaTaO_3$ it is mainly the permittivity maximum that changes, shifting toward lower temperatures, while being depressed and broadened over a wide temperature range. Samples with morphotropic phase compositions exhibit intense dielectric relaxations in the room-temperature region, which correspond also to high values of permittivity. These characteristics make samples from the $Na_{0.5}Bi_{0.5}TiO_3$ and $NaTaO_3$ system attractive for pressure-tunable application, e.g., for pressure sensors.

INTRODUCTION

Sodium bismuth titanate, $Na_{0.5}Bi_{0.5}TiO_3$, is a complex perovskite that was first synthesized by Smolenskii and Agranovskaya in 1959[1]. It forms solid solutions with many other oxide perovskites, e.g., $BaTiO_3$,[2] $SrTiO_3$,[3] $PbTiO_3$,[4] $BaZrO_3$[5], and $Li_{3x}La_{(2/3)-x}\square_{(1/3)-2x}TiO_3$[6]. These systems were investigated with respect to their phase-transition behavior or electromechanical properties in relation to the eventual morphotropic phase boundaries. The results of the studies revealed that ferroelectrics tend to stabilize the room-temperature ferroelectric phase $Na_{0.5}Bi_{0.5}TiO_3$, while incipient ferroelectrics shift the $Na_{0.5}Bi_{0.5}TiO_3$ phase transitions toward lower temperatures. In the latter case increased dielectric relaxations across wide frequency and temperature ranges were formed.

$NaTaO_3$ possesses a negative temperature coefficient of permittivity and may act in a similar way to incipient ferroelectrics.[7] The addition of $NaTaO_3$ shifts the $Na_{0.5}Bi_{0.5}TiO_3$ phase transitions toward lower temperatures and thus has a remarkable influence on its electrical properties.[8] Samples from the $Na_{0.5}Bi_{0.5}TiO_3$–$NaTaO_3$ system that exhibit a dielectric maximum in the room-temperature region have a diminished coercive field, while their remanent polarization is still relatively high. For these samples we anticipate that they show enhanced influence of the mechanical field on their dielectric properties and are therefore attractive for various pressure sensors. Similar behavior was observed during studies of the axial-pressure effect on the permittivity of $Na_{0.5}Bi_{0.5}TiO_3$ and other $Na_{0.5}Bi_{0.5}TiO_3$-based materials.[9,10] These studies revealed that the effect of the axial pressure is most pronounced at the dielectric maximum and that it is strongly reduced at temperatures below the maximum, which was related to variations of the samples` ferroelectric properties.

The aim of our study was to prepare various samples from the $Na_{0.5}Bi_{0.5}TiO_3$–$NaTaO_3$ solid solution, especially those that exhibit a dielectric maximum in the room-temperature region. For these samples we correlated their structure with the dielectric properties and estimated their potential applicability for pressure-tunable devices.

EXPERIMENTAL

Ceramic samples from the $Na_{0.5}Bi_{0.5}TiO_3$–$NaTaO_3$ system were prepared using the solid-state reaction method. Stoichiometric amounts of reagent-grade powders (Alfa Aesar, Karlsruhe, Germany) of Na_2CO_3 (99.997%), Bi_2O_3 (99.975%), TiO_2 (99.8%), and Ta_2O_5 (99.993%) were weighed and mixed in an agate mortar under ethanol. The Na_2CO_3 powder was dried for 2 h at 200 °C before weighing in order to remove any water. The mixed powders were dried, uniaxially pressed into pellets under a pressure of 100 MPa, and calcined in air at 750 °C and 850 °C for 10 h with intermediate cooling and grinding. The calcined samples were milled for 1 h at 200 rpm in a planetary mill using 3-mm yttria-stabilized-zirconia balls and ethanol media. Then, the powders were dried, uniaxially pressed into pellets at 100 MPa and sintered for 5 h in air. The optimal sintering temperatures of the individual compositions were experimentally determined from preliminary firings and range between 1150 and 1320°C.

X-ray powder diffraction (XRD) was used for the phase identification of the samples after each firing. Room-temperature XRD patterns were recorded in the 2θ range from 10°–70° using CuKα radiation (Bruker AXS D4 Endeavor, Karlsruhe, Germany). When more accurate data were required, we used a powder diffractometer with Cu Kα radiation in configuration with Johannson's monochromator to remove the Cu $K\alpha_2$ radiation (PANalytical X'Pert PRO, Almelo, The Netherlands). The XRD patterns were inspected using the EVA software package (Bruker AXS, Karsruhe, Germany) to identify the phases present.

The microstructures of polished and etched sintered samples were investigated using a scanning electron microscope (SEM, Jeol JXA 840A and JSM 5800, Tokyo, Japan). The chemical compositions of the samples were determined with electron-probe microanalysis using an energy-dispersive X-ray spectrometer (EDS, Oxford-Link Isis 300, Oxford Instruments, Oxford, U.K.). The temperature-dependent dielectric measurements were made using an LCR meter (Agilent 4284A, Santa Clara, California, U.S.), a home-made furnace and a temperature chamber (Delta Design 9039, San Diego, California, U.S.) at frequencies from 1 kHz to 1 MHz during heating from –170 to 550 °C. A silver paste was fired onto the samples at 550 °C for 15 min to serve as an electrode.

RESULTS AND DISCUSSION

The x-ray analysis of the samples after the first calcination (10 h at 750 °C) showed the presence of a perovskite matrix phase as well as traces of other crystalline phases, mainly $Bi_4Ti_3O_{12}$ (Figure 1). After the second calcination (10 h at 850 °C) and sintering, XRD analysis revealed that the samples are single phase. Such findings indicate that the solid solutions between $Na_{0.5}Bi_{0.5}TiO_3$ and $NaTaO_3$ can form across the whole concentration range. On the basis of the structures of $Na_{0.5}Bi_{0.5}TiO_3$[11] (rhombohedral perovskite) and $NaTaO_3$[12] (orthorhombic perovskite), the most probable mechanism of substitution involves the exchange of the $(Na,Bi)^{2+}$ pseudo-divalent cation on the A site of the perovskite ABO_3 structure of $Na_{0.5}Bi_{0.5}TiO_3$ with a Na^+ cation, while Ta^{5+} ions substitute for the Ti^{4+} ions on the B site of the perovskite structure.

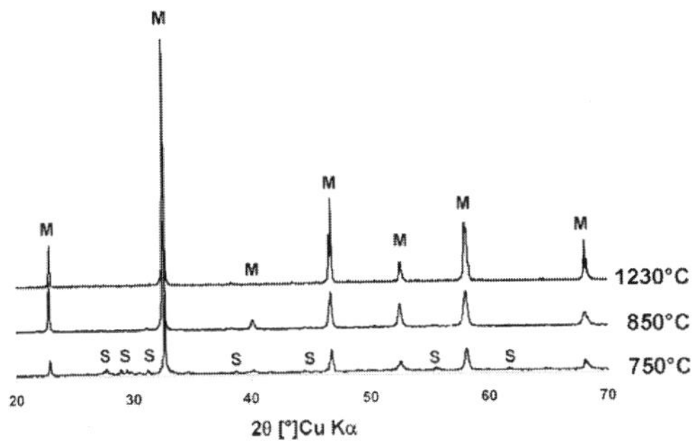

Figure 1: Diffraction patterns for sample with 50 mol% of $NaTaO_3$ after different firings. M denotes the matrix, while S relates to the secondary phase.

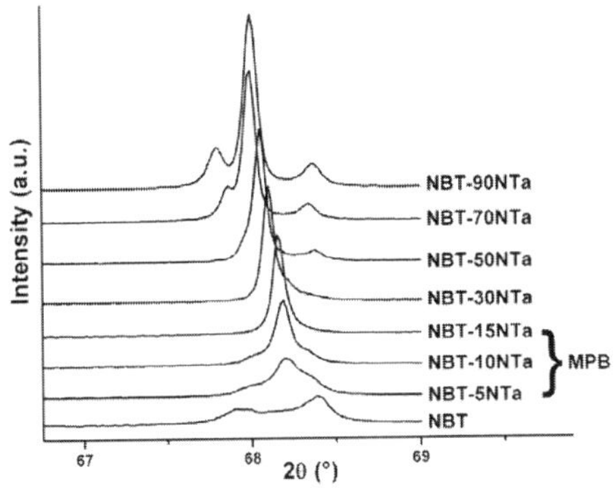

Figure 2: Diffraction peaks at $2\theta \approx 68°$ for sintered samples from $Na_{0.5}Bi_{0.5}TiO_3$–(mol%)$NaTaO_3$ solid solution. The data for pure $Na_{0.5}Bi_{0.5}TiO_3$ are taken from reference[13] (sintered for 15 h at 1150 °C). NBT, NT and MPB correspond to $Na_{0.5}Bi_{0.5}TiO_3$, $NaTaO_3$ and MPB, respectively.

Due to the small differences in the radii of the A- and B-site ions between $Na_{0.5}Bi_{0.5}TiO_3$ and $NaTaO_3$ only a slight shift in the diffraction peaks (0.4°) was observed. The unit-cell size slightly increases across the solid-solution series from $Na_{0.5}Bi_{0.5}TiO_3$ toward $NaTaO_3$, as shown by the shift of the diffraction peaks toward smaller 2θ values in Figure 2. At room temperature the $Na_{0.5}Bi_{0.5}TiO_3$ and $NaTaO_3$ symmetries are rhombohedral and orthorhombic, respectively. From the samples across the solid solution we observed that the transition between the crystal symmetries is diffuse and it is therefore difficult to determine the boundary composition separating the rhombohedral and the orthorhombic phases. Moreover, a detailed XRD scan using monochromatic $K\alpha_1$ radiation revealed the presence of a morphotropic phase composition in samples with 5, 10 and 15 mol% of $NaTaO_3$. As can be seen in Figure 2, the XRD patterns of these compositions resemble the diffraction lines of rhombohedral and orthorhombic symmetries.

The SEM analysis of the microstructures of the sintered samples showed single-phase ceramics for all the prepared compositions. Furthermore, the EDS analysis exhibited no deviations from the nominal compositions within the experimental error of the method. These results are in accordance with the XRD analysis and thus confirm the existence of the solid-solution. Typical microstructures of the prepared samples are shown in Figure 3. The density of the samples was higher than 95% of the theoretical density. With an increasing content of $NaTaO_3$ the sintering temperature increases and the average grain size decreases, as summarized in Figure 4.

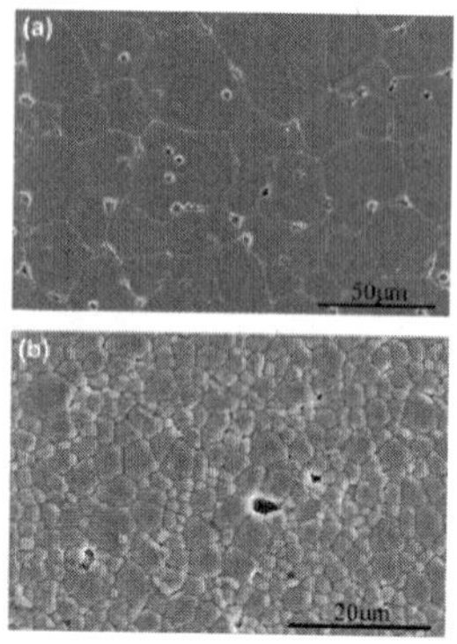

Figure 3: Microstructures of the thermally etched samples with 5 (a) and 90 (b) mol% of $NaTaO_3$ observed using a scanning electron microscope.

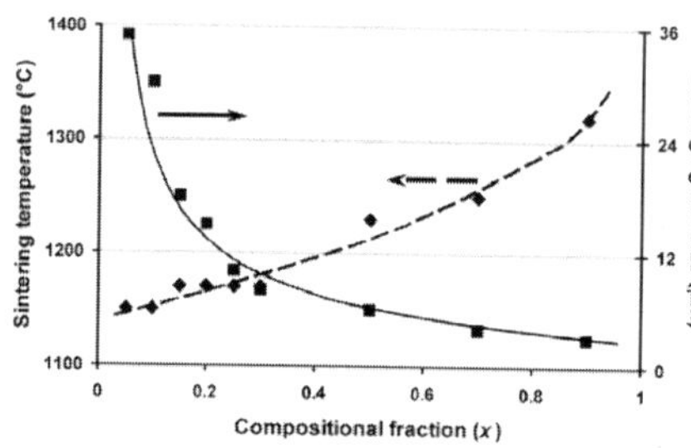

Figure 4: Sintering temperature and average grain size (estimated from the microstructures) of the $(1-x)(Na_{0.5}Bi_{0.5})TiO_3$–$xNaTaO_3$ solid-solution series as a function of the compositional fraction (x).

The temperature dependence of the relative permittivity and the dielectric losses for the investigated samples are shown in Figure 5. An increase in the $NaTaO_3$ content leads to the following changes in the dielectric response:

- permittivity maximum decreases and broadens,
- temperatures of the permittivity maximum and the hump decrease,
- room-temperature dielectric losses decrease.

In the samples with 5 and 10 mol% of $NaTaO_3$ the characteristics of the dielectric response are similar to those of pure $Na_{0.5}Bi_{0.5}TiO_3$. For these samples the permittivity maximum and the hump can be clearly distinguished from each other. Samples with larger additions of $NaTaO_3$ have a strongly depressed permittivity maximum and thus they show a broad plateau with a small temperature dependence of permittivity. As the concentration of the $NaTaO_3$ increases the room-temperature permittivity of the samples first increases up to 900, which is the value for the sample with 15 mol% of $NaTaO_3$. However, for samples with a higher concentration of additive the value is again decreased, as presented in Figure 6. On the other hand, Figure 5 also shows that the temperature of the permittivity maximum monotonously decreases with the addition of $NaTaO_3$, confirming that in the investigated system $NaTaO_3$ acts as an incipient ferroelectric.

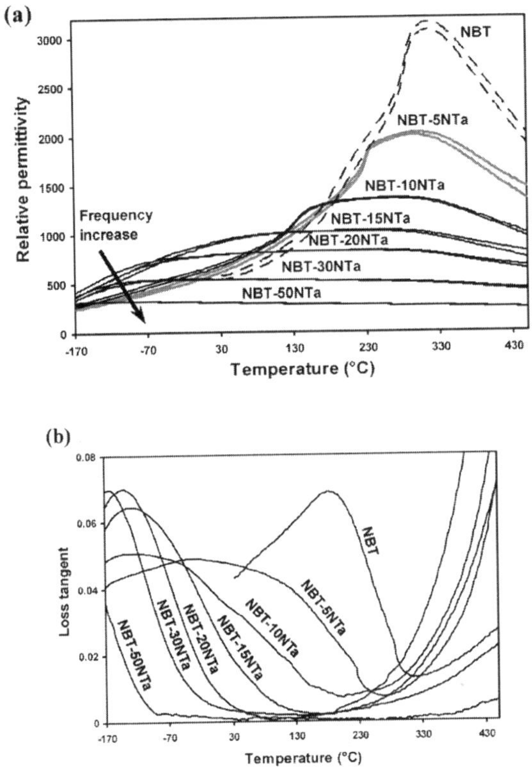

Figure 5: Temperature dependence of relative permittivity (a) and dielectric losses (b) of samples from the $Na_{0.5}Bi_{0.5}TiO_3$–(mol%)$NaTaO_3$ solid-solution series. Values of the relative permittivity were measured at frequencies between 100 kHz and 1 MHz, while the dielectric losses correspond to measurements at 1 MHz.

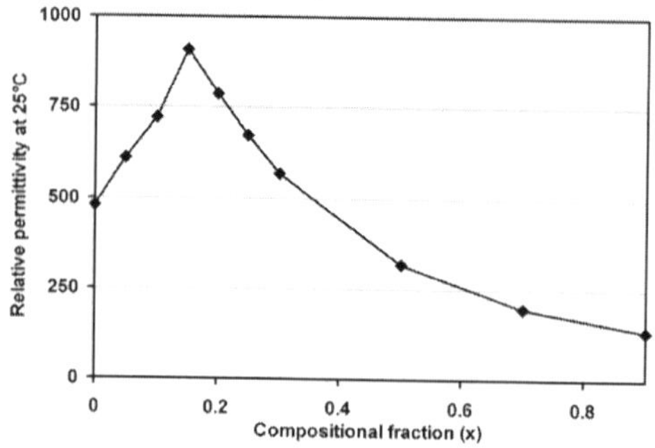

Figure 6: Room-temperature permittivity of the (1-x)($Na_{0.5}Bi_{0.5}$)TiO_3–$x$$NaTaO_3$ solid-solution series as a function of the compositional fraction (x).

CONCLUSIONS

The results revealed that the solid solutions between $Na_{0.5}Bi_{0.5}TiO_3$ and $NaTaO_3$ form across the whole concentration range. Additionally, morphotropic phase compositions were found for samples with 5, 10 and 15 mol% of $NaTaO_3$. With increasing content of $NaTaO_3$ the dielectric anomalies shift toward lower temperatures, confirming that $NaTaO_3$ acts as an incipient ferroelectric. In addition, the permittivity maximum is depressed and broadens over a wide temperature range. The samples with morphotropic phase compositions exhibit intense dielectric relaxations in the room-temperature region, which correspond also to high values of permittivity. Therefore, the enhanced influence of the mechanical field on their dielectric properties is expected for samples with $NaTaO_3$ around 15 mol%, making them attractive for pressure-tunable applications.

REFERENCES

[1] G. A. Smolenskii and A. I. Agranovskaya, "Dielectric Polarization of a Number of Complex compounds," *Sov. Phys. Solid State*, **1** 1429–1437 (1960).

[2] T. Takenaka, K. Maruyama, and K. Sakata, "($Bi_{1/2}Na_{1/2}$)TiO_3–$BaTiO_3$ System for Lead-Free Piezoelectric Ceramics," *Jpn. J. Appl. Phys.*, **30** 2236–9 (1991).

[3] J. R. Gomah-Pettry, A. N. Salak, P. Marchet, V. M. Ferreira, and J. P. Mercurio, "Ferroelectric Relaxor Behaviour of $Na_{0.5}Bi_{0.5}TiO_3$–$SrTiO_3$ Ceramics," *Phys. Stat. Sol. B*, **241** [8] 1949–56 (2004).

[4] J. K. Lee, K. S. Hong, C. K. Kim, and S. E. Park, "Phase Transformations and Dielectric Properties in A-site Ion Substituted ($Na_{1/2}Bi_{1/2}$)TiO_3 Ceramics (A=Pb and Sr)," *J. Appl. Phys.*, **91** [7] 4539–42 (2002).

[5] S. A. Sheets, A. N. Soukhojak, N. Ohashi, and Y. M. Chiang, "Relaxor Single Crystal in the $(Na_{1/2}Bi_{1/2})_{1-x}Ba_xZr_yTi_{1-y}O_3$ System Exhibiting High Electrostrictive Strain," *J. Appl. Phys.*, **90** [10] 5287–95 (2001).

[6] M. Spreitzer, B. Jančar and D. Suvorov, "Phase Relations in the $Na_{0.5}Bi_{0.5}TiO_3$–$Li_{3x}La_{(2/3)-x\ (1/3)-2x}TiO_3$ System," *Int. J. App. Ceram. Tech.*, in press.

[7] H. Iwasaki and T. Ikeda, "Studies on the System $Na(Nb_{1-x}Ta_x)O_3$," *J. Phys. Soc. Jap.*, **18** [2] 157–63 (1963).

[8] J. König, B. Jančar and D. Suvorov, "New $Na_{0.5}Bi_{0.5}TiO_3$–$NaTaO_3$-Based Perovskite Ceramics," *J. Am. Ceram. Soc.*, **90** [11] 3621–3627 (2007).

[9] J. Suchanicz and T. V. Kruzina, "The Effect of Axial Pressure on Domain State and Dielectric Properties of $Na_{0.5}Bi_{0.5}TiO_3$ and Related Materials," *Ferroelectrics*, **317** 109–13 (2005).

[10] J. Suchanicz, J. P. Mercurio, P. Marchet, and T. V. Kruzina, "Axial Pressure Influence on Dielectric and Ferroelectric Properties of $Na_{0.5}Bi_{0.5}TiO_3$ Ceramic," *Phys. Stat. Sol. B,* **225** [2] 459–66 (2001).

[11] G. O. Jones and P. A. Thomas, "Investigation of the Structure and Phase Transitions in the Novel A-site Substituted Distorted Perovskite Compound $Na_{0.5}Bi_{0.5}TiO_3$," *Acta Crystallogr.*, **B58** 168–78 (2002).

[12] B. J. Kennedy, A. K. Prodjosantoso, and C. J. Howard, "Powder Neutron Diffraction Study of the High Temperature Phase Transitions in $NaTaO_3$," *J. Phys.: Condens. Matter*, **11** 6319–27 (1999).

[13] M. Spreitzer, M. Valant, and D. Suvorov, "Sodium Deficiency in $Na_{0.5}Bi_{0.5}TiO_3$," *J. Mater. Chem.*, **17** 1–9 (2007).

PIEZOELECTRIC BEHAVIOUR OF THE BLENDED SYSTEMS (NYLON 6/NYLON 11)

S. A. Pande
G.H. Raisoni College of Engineering, Nagpur – 440 016, Maharashtra, India
D. S. Kelkar
Dept. of Physics, Institute of Science, Nagpur – 440 001, Maharashtra, India
D. R. Peshwe
Material Engineering Centre, Visvesvaraya National Institute of Technology, Nagpur - 440 011, Maharashtra, India

ABSTRACT

There is great importance for the study of combination of an engineering plastic polyamide and the best commodity product Nylon-6, which could lead to new blends with better piezoelectric properties. Keeping commercial approach in view Nylon 6 and Nylon 11 blends have been prepared. Compositions of two different proportions 90 :10 and 10:90 were developed.

The blended samples were poled using Corona discharge technique with poling field of 200 kV/cm and 350 kV/cm. The piezoelectric coefficients (d_{33}), as a function of temperature for blended samples were measured up to 100°C with Piezometer. The study suggests that due to mixing of small quantity of Nylon 11 in the Nylon 6 (10:90) the piezoelectric response has improved considerably despite of the fact that Nylon 6 is not a piezoelectric material. The study indicates that even blending of small quantity (10%) of Nylon 11 in the (90%) Nylon 6 generates good piezoelectric response. The measured piezoelectric response of such compositions is almost equal to the pure Nylon 11.

INTRODUCTION

Recently the odd-numbered nylons have gained the premier importance in the field of investigation due to its extra-ordinary capability to impart piezoelectricity. In the past three decades polymer blends technology has been one of the most investigated areas in Polymer Science. With increased academic and industrial research interest, the application of polymer blends technology to commercial utility has grown significantly.

In 1980, Newman et al[1] while investigating piezoelectric response of Nylon 11 found that the piezoelectric strain coefficient, $d_{31} \approx 3$ pC/N for poled Nylon 11 film. Scheinbeim[2] for melt quenched and poled Nylon 11 films reported rapid increase in piezoelectric coefficient with increasing temperature. Mathur et al[3] reported that the piezoelectric strain and stress coefficients (d_{31} and e_{31}) showed hysteresis behaviour.

Lee et al[4-5] for the first time reported that melt-quenched, cold-drawn, and then poled Nylon 11 films exhibited an electric displacement D versus electric field E hysteresis loop, which clearly indicated the ferroelectric behaviour. It was also reported that melt-quenched undrawn Nylon 11 films exhibits clear D-E hysteresis behaviour[6-7]. Wu S. L. et al[8] studied the effect of melt-quenched Nylon 11 films, uniaxially drawn, and observed that as the draw ratio of the films increases, the piezoelectric strain coefficient d_{31} at 25°C remains unchanged and the stress coefficient e_{31} increases linearly, with residual polarization (Pr) under dry conditions. Biao-bing Wang et al[9] studied the characterization of Nylon 6/11 copolymer and reported that intrinsic viscosity of Nylon 6/11 copolymer was influenced by the caprolactam content as well as copolymerization time under vacuum. However, the incorporation of caprolactam into Nylon 11 chains did not transforms the crystal phase of Nylon 11.

Though Nylon 11 is a good piezoelectric material, and also possess other advantages over the similar materials, still it could not be used in the industrial application due to its very high cost. Moreover less attention has been paid to study the piezoelectric properties of Nylon-11 blends. Therefore, much work has not been published on this aspect. It was therefore considered appropriate to study the variation in the piezoelectric response of blends of Nylon-6/ Nylon-11 in comparison with the individual Nylon-6 and Nylon-11.

In the present work, blends were obtained in the proportion 90/10 and 10/90 for Nylon-6/ Nylon-11 by melt casting..

EXPERIMENTAL TECHNIQUE

Materials

Nylon-11 (Sigma Aldrich CHEMIE GmbH, Germany) and Nylon 6 (M/s Nirlon Synthetic Fibers & Chemicals Mumbai, India) in the form of beads were used in the preparation of samples. The blends of Nylon-6/ Nylon-11 were prepared in proportion (90:10 & 10:90). The blending was achieved through single screw excruder with chopper. The blend granules were then fed to hot press molding machine to obtain circular samples of nominal thickness of 1 mm. All the samples were stored in the desiccator.

Wide Angle X-Ray Diffraction (WAXD):

WAXD is a widely used technique to study the crystallinity of a polymer. Phillips difffractometer with CuKα radiation was used in this study. The wide angle x-ray diffraction of pure and blend of Nylon 11/Nylon 6 were recorded at room temperature with 1^0/min scan over a range of 2θ from 15^0 to 40^0.

Scanning Electron Microscopy (SEM):

It is one of the most important techniques used for the investigation of morphology of polymers. During the present work, the morphology was investigated using Leo 435 VP 501B Philips SEM.

The samples to be examined were made conductive by depositing a very thin layer of gold of thickness 800 nm on the surface. Initially large area of the samples was observed at low magnification and then the selected area was studied at higher magnification.

Poling and Measurements:

Gold electrodes were sputtered on one surface of these samples under high vacuum to establish electrical contact. These samples were then poled using corona discharge. A voltage gradient of 200 kV/cm and 350 kV/cm were applied across the samples for 1 hour at various temperatures. After poling, the samples were allowed to relax at room temperature under field and then piezoelectric strain coefficient d_{33} was measured at room temperature using Piezometer systems PM-25.

The piezoelectric strain coefficient d_{31} was calculated from the measured d_{33} values using the following relation[10] which holds only in case of the low poling field.

$$d_{31} = d_{33}/\,3 \qquad \text{(i)}$$

The piezoelectric stress coefficient e_{31} can be calculated [13]

$$e_{31} = d_{31} * Y \qquad \text{(ii)}$$

Where Y = Young's modulus.

The electromechanical coupling coefficient, k_{31} can be obtained from the following equation[11]

$$k_{31} = d_{31} \sqrt{\frac{Y}{\varepsilon_0 \, \varepsilon'}} \qquad \text{(iii)}$$

Where Y = Young's Modulus, ε_0 = permittivity of free space and ε' is the dielectric constant at room temperature at 1KHz.

Thus using the above relations the piezoelectric coefficients d_{31}, e_{31}, k_{31} were calculated at two different poling fields and at various temperatures.

Piezoelectric Coefficient (d_{33})

Figure [1 (a) and (b)] show the temperature dependence of piezoelectric coefficient (d_{33}) at two different poling fields for pure and blended Nylon 11.

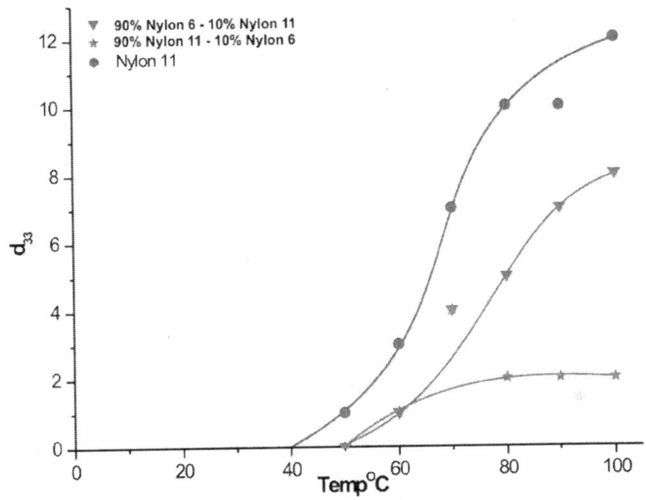

Figure 1(a) : Temperature dependence of the piezoelectric strain coefficient d_{33} at poling field 200 kV/cm

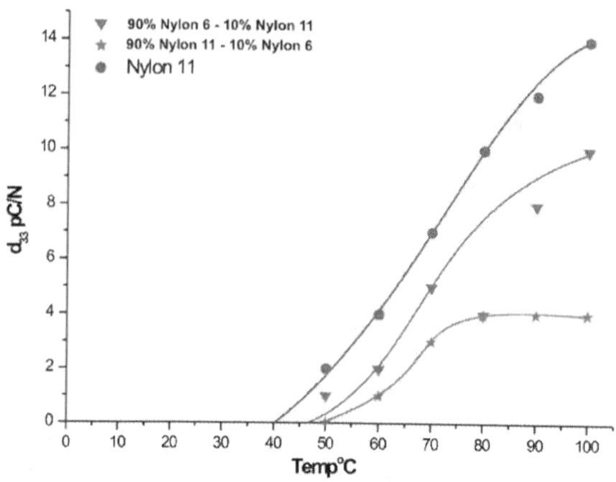

Figure 1(b) : Temperature dependence of the piezoelectric strain coefficient d_{33} at poling field 350 kV/cm

From the figures, it can be observed that d_{33} is measurable when the poling temperature exceeds ~ 45°C. At low poling field i.e. at 200 kV/cm, the strain coefficient d_{33} increases rapidly with increasing temperature nearly up to 70°C. In case of pure Nylon 11 the maximum value for piezoelectric coefficient (d_{33} ~ 12 pC/N) has been observed at 100°C. As Nylon 6 does not exhibit piezoelectric response the value of d_{33} is found to be zero.

Similar results were obtained when the poling field increased to 350 kV/cm, wherein the maximum value for piezoelectric coefficient obtained for Nylon 11 was d_{33} ~ 14 pC/N. The co-relation between the various piezoelectric coefficients has already been discussed earlier and by using the above referred equations piezoelectric coefficients d_{31}, e_{31} were calculated. The detailed comparison in respect of variation in piezoelectric strain (d_{31}) and stress coefficient (e_{31}) for 200 kV/cm and 350 kV/cm is given in Table I.

Table I: Values of d_{31} and e_{31} for blended compositions at two different poling fields.

Sr. No.	Blend Compositions		Samples	Poling fields at 100°C			
				200 kV/cm		350 kV/cm	
	Nylon 6	Nylon 11		d_{31} (pC/N)	e_{31} (mC/m^2)	d_{31} (pC/N)	e_{31} (mC/m^2)
1	-	100%	Pure Nylon 11	4.00	5.11	4.67	5.97
2	90%	10%	Blend of Nylon 6/Nylon11	2.67	2.35	3.33	2.93
3.	10%	90%	Blend of Nylon 6/Nylon 11	0.67	0.42	1.33	0.84

The emphasis was laid to calculate the d_{31} and e_{31} coefficients because the available literature survey, given above has discussed only about piezoelectric strain coefficient (d_{31}) of Nylon 11 and the data of d_{33} of Nylon 11 was not available for comparison. Therefore in the present study though d_{33} coefficient was measured, it cannot be co-related. Hence in order to correlate the obtained results with the available literature, d_{31} was calculated from d_{33} coefficient using equation (i).

Electromechanical coupling coefficient, k_{31}, is obtained from equation (iii), for the pure and blended compositions. The values of k_{31} for poling field of 200 kV/cm and 350 kV/cm has been tabulated in Table II.

Table II: Values of k_{31} at two different poling fields.

Sr. No.	Blend Compositions		Samples	Poling Field at 100^{0}C	
	Nylon 6	Nylon 11		200 kV/cm	350 kV/cm
				k_{31}	k_{31}
1.	-	100%	Pure Nylon 11	0.0189	0.0220
2.	90%	10%	Blend of Nylon 6/Nylon 11	0.0098	0.0122
3.	10%	90%	Blend of Nylon 6/Nylon 11	0.0021	0.0042

It is interesting to note that Nylon 6, a centro-symmetric material, does not exhibit a piezoelectric response. However its blend with Nylon 11 shows considerable piezoelectric response.

Piezoelectric properties of polymers are closely related to the chemical and crystalline structure of materials and the dipole density in the crystalline unit cell. It also appears that piezoelectric activity simply results from the aggregate properties of polar crystallites with preferred dipole orientation in an amorphous matrix. The all-transconformation of the odd Nylon molecules such as Nylon 9 and Nylon 11 gives a large dipole moment perpendicular to the chain axis and all dipole moments are aligned in the same direction. So the odd Nylons can have a spontaneous polarization in the unit cell of the crystalline phase, and thus the odd Nylons are good piezoelectric materials.

Researchers[12-15] have reported that under equivalent conditions, the piezoelectric constants of Nylon 11 obtained from γ structure are higher than those of α structure. It has been proposed that the field induced dipole alignment might more effectively occur in the γ structure than in ordinary α structure. Similarly, G Wu et al[16] reported that under equivalent poling conditions Nylon 11 exhibits a larger piezoelectric response for the melt quenched specimen consisting of the mixture of α and γ forms than the pure α or pure γ forms.

It was established that the piezoelectricity in the material is dependent upon the crystal structure of the material and therefore it is necessary to ascertain the crystal structure of pure and blended samples. The structural analysis of pure and blended Nylon 11 using WAXD analysis, wherein the x-ray scan of unpoled samples only have been recorded. It has already been reported[17] that there are negligible changes due to poling in x-ray.

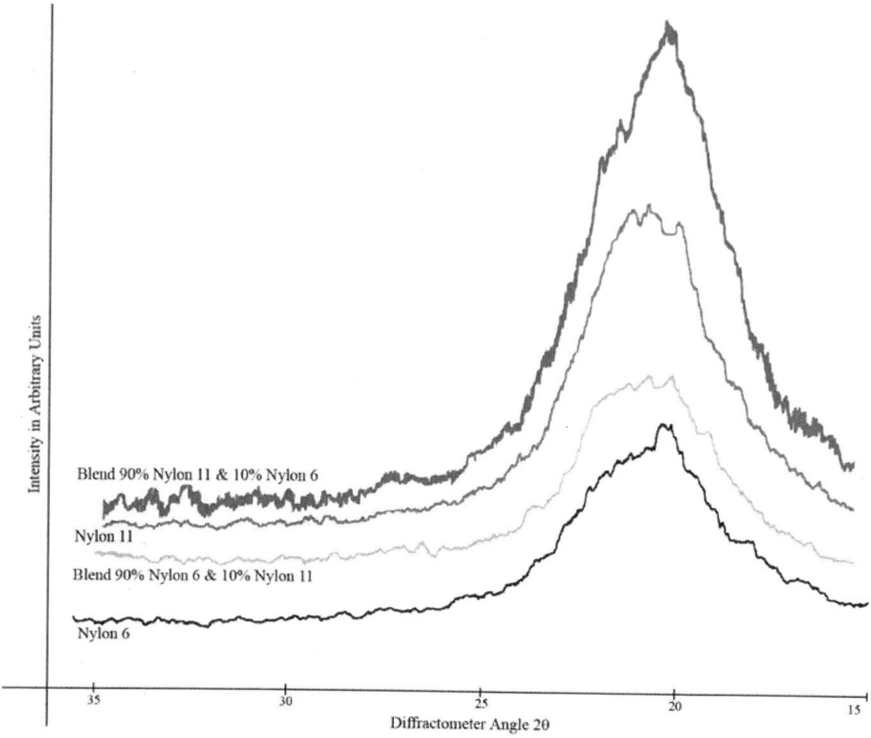

Figure 2: WAXD scan of pure Nylon 6, pure Nylon 11 & Blends

The XRD scan for the pure individual polymers i.e. Nylon 11 and Nylon 6 are shown in Figure 2. From the figure it is found that three distinct and prominent peaks in Nylon 11 were appearing at $2\theta = 20^\circ$, 20.8° corresponding to α form and the one at 21.8° corresponding to γ form, while *Nylon 6* portrayed α form at $2\theta = 20.8^\circ$ and γ form at $2\theta = 21.2^\circ$.

However, when these two polymers have been blended with each other in the proportions (90/10 and 10/90), the XRD pattern differs to a large extent (Figure 2). In the sample 90% Nylon 6 and 10% Nylon 11, well defined and intense peaks are observed at $2\theta = 20^\circ$ (α-form) and 21° & 21.8° (γ-form) respectively. And an extra broad peak appearing at 22.4° indicates formation of γ form. Similarly in sample 10% Nylon 6 and 90% Nylon 11, well defined and intense peaks are observed at $2\theta = 20.8^\circ$ (α-form) and 22.2° and 22.8° (γ-form) as compared to pure Nylon 11.

Thus from overall XRD analysis we conclude that crystal modification has taken place and altogether different structure has been formed, which does not resembles with the individual polymers. The appearance of extra small peak might be due to formation of a few small crystals having different crystal modification. Similarly, the SEM micrographs (Figure 3) of blended sample predict totally different type of behaviour which indicates that the complete structure has been disturbed and a complex structure has been formed.

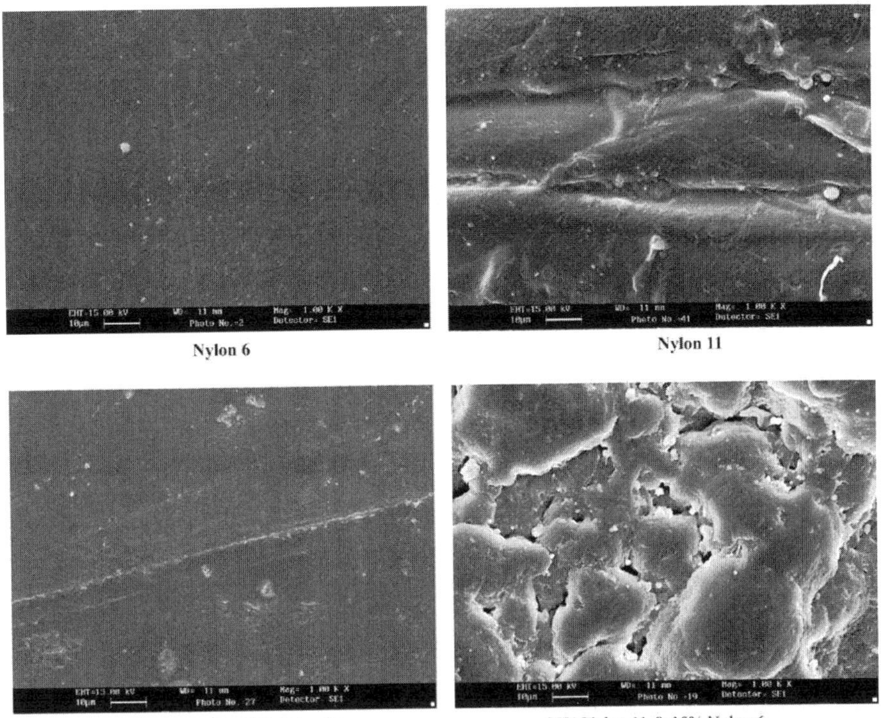

Nylon 6 | Nylon 11

90% Nylon 6 & 10% Nylon 11 | 90% Nylon 11 & 10% Nylon 6

Figure 3: SEM Micrographs of pure and blended samples of Nylon 11 and Nylon 6

Nylon 11 is having the higher dipole concentration and non-centrosymmetric unit cells, while Nylon 6 though containing the dipoles, its resultant dipole concentration is zero due to the centro-symmetric unit cells. However, after the addition of the Nylon 6 in the Nylon 11 matrix the blended composition exhibits piezoelectric response, though not more than the pure Nylon 11. The reason for this can be attributed to the resultant dipole moment. In pure Nylon 11, since the direction of the dipole moment is the same for all dipoles, the material exhibits the resultant dipole moment.

Addition of Nylon 6 in Nylon 11 in proportion (90:10 & 10:90), it is more likely that hydrogen bonds will form among Nylon 11 and Nylon 6 such a unit cell then can have a resultant dipole moment which will increase with application of poling field . Thus blended samples rich in Nylon 6 exhibits piezoelectric properties.

Similar situation can also arise in blend systems 10% Nylon 6 and 90% Nylon 11. In this case presence of Nylon 6 chains in Nylon 11 matrix disturbed the original dipole alignment of Nylon 11 and hence reduction in piezoelectric response is obtained.

CONCLUSION

The piezoelectric response of pure and blended samples was studied at two different poling fields. The blending of Nylon 11 (non-centrosymmetric) by Nylon 6 (centrosymmetric), a better piezoelectric response was observed. Thus, addition of 90% Nylon 6 to 10% Nylon 11 is sufficient to impart piezoelectric properties to blends. The results are interpreted on the basis of structural changes in the blended samples.

ACKNOWLEDGEMENT

The authors are thankful to Advanced Ceramics Laboratory, Department of Applied Physics, IIT Delhi, AIIMS, New Delhi and Naval Material Research laboratory (NMRL), Ambernath, Mumbai for providing the facilities and support to carry out the measurements of the samples.

REFERENCES

1. Newman B. A., Chen P., Pae K. D., Scheinbeim J. I., J. Appl. Phys., 51, 5161, 1980.
2. Scheinbeim J. I., J. Appl. Phys., 52, 5939, 1981.
3. Mathur S C., Scheinbeim J. I., Newman B A., J. Appl. Phys., 56, 2419, 1984;
4. Lee J.W., Takase Y., Newman B.A., Scheinbeim J.I., J. Polym. Sci. Part B: Polymer Phys., 29, 273, 1991.
5. Scheinbeim J.I., Lee J. W., Newman B A., Macromolecules, 25, 3729, 1992.
6. Mei B.Z., Scheinbeim J.I., Newman B A., Ferroelectrics, 144, 51, 1993.
7. Mei B.Z., Scheinbeim J.I., Newman B A., Ferroelectrics, 171, 177, 1995.
8. Wu S.L., Scheinbeim J.I., Newman B A., J. of Polym Sci., Part B: Polymer Physics, 37, p 2737-2746, 1999.
9. Biao-bing Wang, Guo-sheng Hu, Xin Zhao, Feng-zhen Gao, Materials Letters, 60, 21-22, 2715-2717, 2006.
10. Singer K.D., Kuzyk M. and Sohn J.E., Second – order nonlinear optical processes in orientionally ordered materials: Relationship between molecular and macroscopic properties, J. Opt. Soc. Am. B4, p 968-976, 1987.
11. Wang J.T., Herbert J.M., Glass A.M., The Appl. of the Ferro. Polymers: Blackie and Sons Ltd. New York., 1988.
12. Newman B.A., Sham T.P. and Pae K.D., J. Appl. Phys. 48, 4092, 1977.
13. Kato K., Furukawa T. and Fukada E., Polym Prep. Jpn, 27, 1800, 1978.
14. Gelfandbein V. and Katz D., Ferroelectrics, 33, 111, 1981.
15. Gelfandbein V. and Katz D., J. Phys. D. Appl. Phys., 15, L115, 1982.
16. Wu G., Yano O., Soen T., Polym. J. 18, 51, 1986.
17. Scheinbeim J.I, Journal Appl. Phys., 52, 10, 1981.

DIELECTRIC PROPERTIES OF $BaTiO3$ DOPED WITH $Er2O3$ AND $Yb2O3$ BASED ON INTERGRANULAR CONTACTS MODEL

V.V. Mitic
Faculty of Electronic Engineering, University of Nis, Nis, Serbia
Institute of Technical Sciences of SASA, Belgrade, Serbia

V. Paunovic, D. Mancic, Lj. Kocic, and Lj. Zivkovic
Faculty of Electronic Engineering, University of Nis, Nis, Serbia

V.B. Pavlovic
Faculty of Agriculture, University of Belgrade, Belgrade, Serbia

ABSTRACT

Taking into account that microstructure properties of barium-titanate based materials, expressed in grain boundary contacts, are of basic importance for electric properties of these material, in this study, the model of intergranular impedance is established using the equivalent electrical scheme characterized by corresponding frequency characteristic. Globally, $BaTiO_3$-ceramics sample is consisted of a huge number of mutually contacted grains which form clusters. For each of them, it is possible to establish the equivalent electrical model and, for defined set of input parameters, using symbolic analysis, obtain the frequency diagram. Realizing the totality of relations between clusters grains groups, their microelectrical schemes and corresponding frequency characteristics, from one side, and global equivalent electrical scheme and corresponding acquired frequency characteristics of $BaTiO_3$-ceramics samples, on the other side, we set a goal of coinciding experimental results with the summing effect of microelectric equivalent schemes. The model is successfully tested on barium titanate ceramics doped with Er_2O_3, Yb_2O_3.

INTRODUCTION

Barium-titanate based materials are widely utilized in the manufacture of multilayer capacitors (MLC-s), thermistors and electro-optic devices [1,2]. Conventional processing of these materials relies on high temperature synthesis of raw materials .Since grain size and distribution considerably affect electrical properties of barium-titanate based materials, correlation of their microstructure and electrical properties has been investigated most extensively by numerous authors [3-5]. Gained results relate to the distribution and the size of grains and pores, to the degree of interior strain within grains boundary, as well as to the very structure of grains boundary.

It has been shown that electrical properties of undoped and doped $BaTiO_3$-ceramics are mainly controlled by barrier structure, domain motion of domain boundaries and the effects of internal stress in the grains [6-7]. Taking into account that microstructure properties of barium-titanate based materials, expressed in grain boundary contacts, are of basic importance for electric properties of these material, in this study, the model of intergranular impedance is established using the equivalent electrical scheme characterized by corresponding frequency characteristic.

EXPERIMENTAL PROCEDURE

The samples were prepared from high purity (>99,98) commercial $BaTiO_3$ powder (MURATA) with [Ba]/[Ti]=1,005 and reagent grade Er_2O_3 and Yb_2O_3 powders (Fluka chemika). Er_2O_3 and Yb_2O_3 dopants were used in the amount to have 0.5 wt% Er or Yb in $BaTiO_3$. Starting powders were ball milled in ethyl alcohol for 24 hours using polypropylene bottle and zirconia balls. After drying at 200°C for several hours, the powders were pressed into disk of 7 mm in diameter and 3 mm in thickness under 120 MPa. The compacts were sintered up to 1380°C in air for four hours. The microstructures of as sintered or chemically etched samples were observed by scanning electron microscope (JEOL-JSM 5300) equipped with energy dispersive x-ray analysis spectrometer (EDS-QX 2000S system). Prior to electrical measurements silver paste was applied on flat surfaces of specimens. Capacitance, dissipation factor and impedance measurements were done using Agilent 4284A precision LCR meter in the frequency range of 20 to 106 Hz. The illustrations of the microstructure simulation, were generated by Mathematica 6.0 software.

RESULTS AND DISSCUSSION

In this study we have considered $BaTiO_3$-ceramics sample as a system with a huge number of mutually contacted grains which form clusters. For each of them, it is possible to establish the equivalent electrical model and, for defined set of input parameters, using symbolic analysis, obtain the frequency diagram. Realizing the totality of relations between clusters grains groups, their microelectrical schemes and corresponding frequency characteristics, from one side, and global equivalent electrical scheme and corresponding acquired frequency characteristics of $BaTiO_3$-ceramics samples, on the other side, we set a goal of coinciding experimental results with the summing effect of microelectric equivalent schemes.

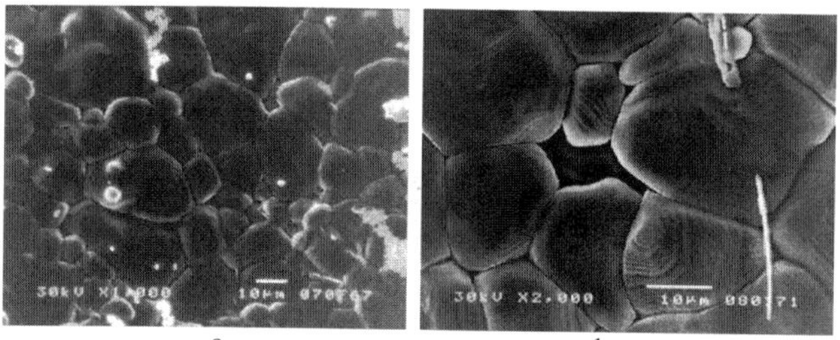

a b

Fig. 1 Microstructures of $BaTiO_3$ doped with a) 0.5 wt% Er_2O_3 b) 0.5 wt% Yb_2O_3.

According to the microstructures we have obtained for $BaTiO_3$ doped with Er_2O_3 and Yb_2O_3 (Fig. 1) it can be concluded that the global impedance of barium-titanate ceramics sample, which contains both resistor and capacity component, can be presented as a "sum" of many clusters of micro-resistors and micro-capacitors connected in tetrahedral lattice. According to our previous results, for the general model of the stereological configuration of $BaTiO_3$, consisting of a cluster of spheres or ellipsoids, density of the sample can be defined by mutual positions of the neighboring grains (Fig. 2).

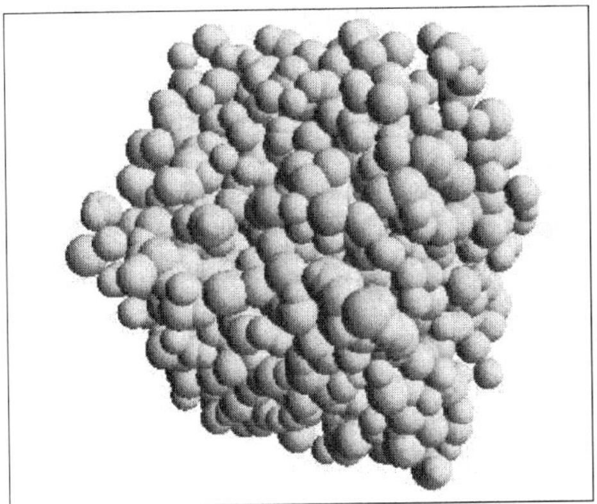

Fig 2. The simulation of the 1000 grains of $BaTiO_3$ obtained from the general model of the stereological configuration.

Developed model gives the distribution of grain contacts through the sample volume. As it can be seen from the typical spatial situation of four grains cluster (Fig 3), the positions of the neighboring grains can be: 1. in touching contact; 2. slightly immerging one into another and 3. not touching each other.

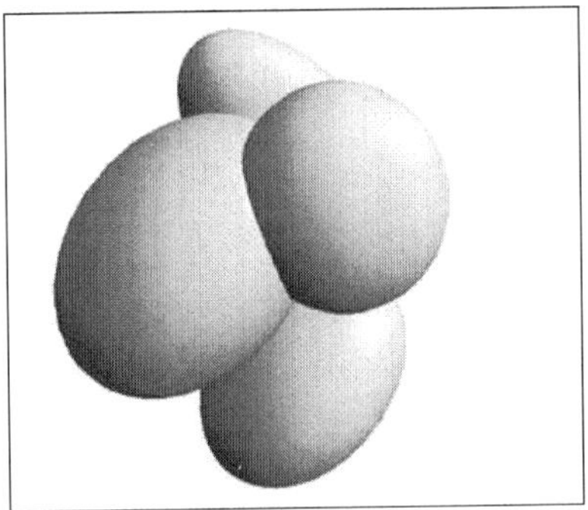

Fig. 3 The positions of neighboring grains for the four grains cluster.

This configuration logically leads to the tetrahedral scheme of mutual electrical influence of $BaTiO_3$ grains. The impedances are at the each edge of tetrahedron, as it is shown on Fig. 4. The vertices (in Fig. 4 displayed as small spheres) are stylized grains, while impedances contain resistance and capacity between two grains.

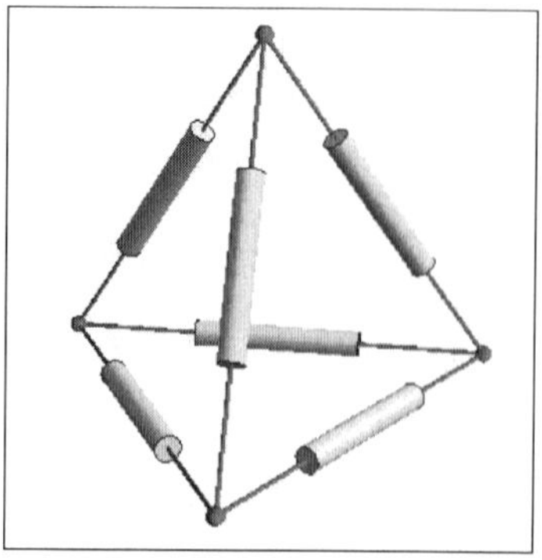

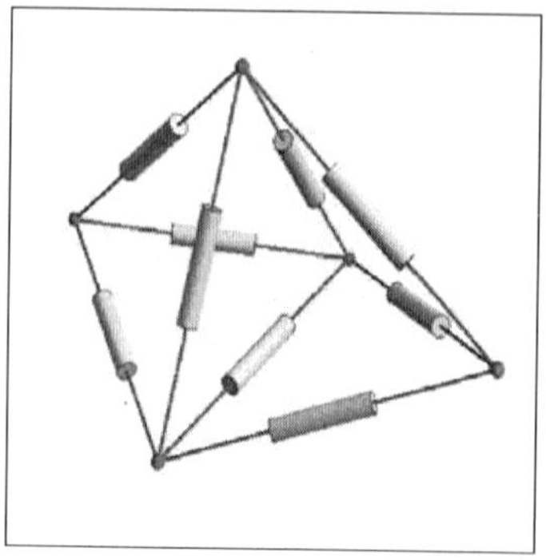

Figure 4. The impedance model of tetrahedron.

Our new approach includes fractal geometry in describing complexity of the spatial distribution of $BaTiO_3$ grains. The best fractal model is a sponge model, or and it is more correct to say, a kind of three –dimensional lacunary set (a set with voids). The structure of tetrahedral influence may be established in each spatial sense, which means that one has a tetrahedral lattice that fills the space.

This kind of fractals sometimes is identified with deterministic constructions like *Cantor set* (in P), *Sierpinski triangle* or *Sierpinski square* (in P^2), with *Sierpinski pyramid* or *Menger sponge* (in P^3), and so on. In our case of electric impedance property,

The *Sierpinski pyramid* (Figure 6) might be a quite adequate paradigm for the first instance inquiry. The starting pyramid $\mathbf{T}_0$ and the first two iterations, shown in Fig. 7, give initial part of the orbit of so called Hutchinson operator W, that is $\mathbf{T}_1 = W(\mathbf{T}_0)$, and $\mathbf{T}_2 = W(\mathbf{T}_1) = W^2(\mathbf{T}_0)$. The limit case, $\mathbf{T} = W^{\infty}(\mathbf{T}_0)$ is an exact fractal set with Hausdorff dimension $D_H = 2$.

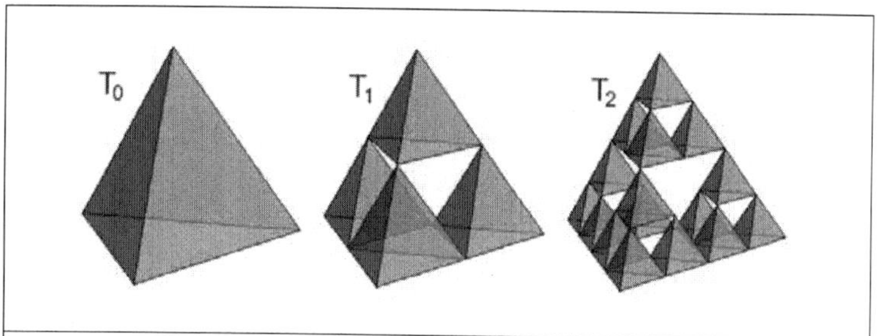

Figure 6. The Sierpinski pyramid

Following this model of Sierpinski pyramid, the induced model of impedances between clusters of ceramics grains is displayed in Fig 7. The task is to calculate equivalent impedance for the pyramid $\mathbf{T}_k$ in function of k.

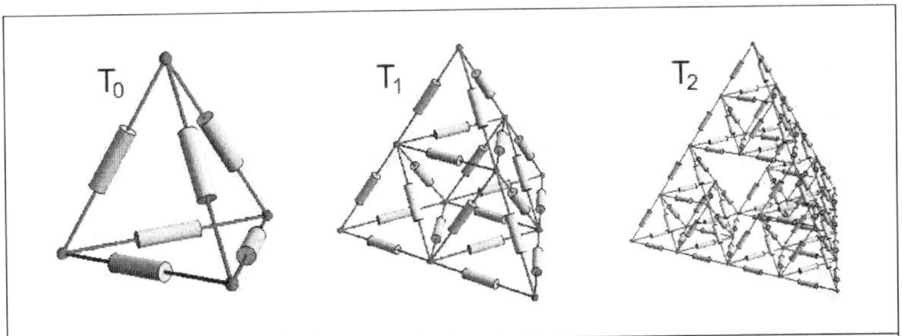

Figure 7. The model of impedances between clusters of ceramics grains.

Some of the impedances in this lattice are missing following the lacunary fractal pattern. According to [8], [9], the fractal dimension of the typical ceramics sample is estimated to be about 2.087. So one may conjecture that fractal dimension of *$BaTiO_3$* –ceramics is slightly below 3. The denser the ceramics the bigger the dimension. The dimension close to 3 means that not many of the impedances lack in the tetrahedral lattice. On the other hand, dimension closer to 2.5 alerts high presence of pores which means that many impedances are to be swept in the model of tetrahedral impedance lattice. What is sure is that $BaTiO_3$ sintered ceramics can be considered as a lacunar fractal rather than a percolation fractal or succolar fractal [10].

In order to calculate equivalent impedance for a wide frequency range, the equivalent electrical circuit for a ceramic material can be introduced as an impedance containing two capacitance C and C_p, an inductance L and a resistance R.

The dominant electrical parameter of our model is the capacitance C. The connection between C and geometrical and/or structural properties can be established by an assumption that the contact region can be viewed as a planar micro-capacitance. Fig.9 shows a geometric model of two spheres in contact, where r_c is the radius of the spherical particles, and x is the neck radius.

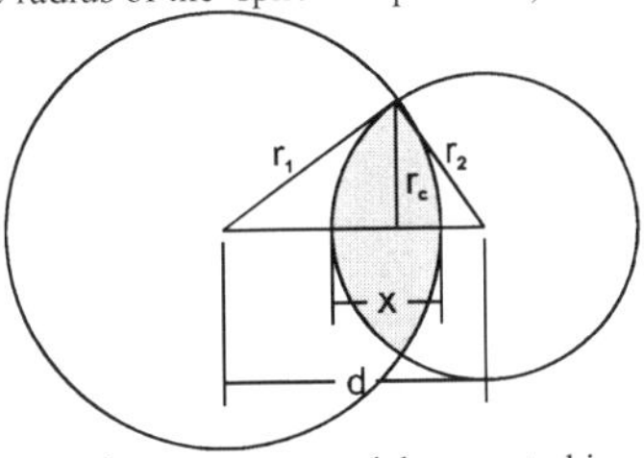

Fig 9 Two-sphere contact model presented in plane section

Since the spherical shape and the circular neck profile are assumed, the contact surface is the circle of the area πx^2. Because of center-to center approach, the dielectric thickness 2h is a function of d (the distance between centers of particles) according to the relatio $2h=2r_c-d$. Therefore, the capacitance can be written as:

$$C = \varepsilon_o \varepsilon_r \alpha \frac{\pi x^2}{2r_c - d} \qquad (d = 2\sqrt{r_c^2 - x^2})$$

where ε_o and ε_r are the diuletric constants of vacuum and the ceramic material, respectively and α is a correction factor obtained by a constructive approach to the fractal structure. Taking into account that by the fractal theory α can be presented as:

$$\alpha = D - D_T$$

where $D \approx 2.08744$ is the fractal (Hausdorff) dimension of intergrain contact surface and $D_T=2$ is topological dimension of the surface, it can be concluded that for $BaTiO_3$ doped ceramics contact surfaces are of low-irregularity which is characterized by the small diference $D - D_T \approx 0.08744$.

Taking this into account, calculations of microcapacitance generated in grains contacts of $BaTiO_3$ doped with Er_2O_3 and Yb_2O_3 have been performed.

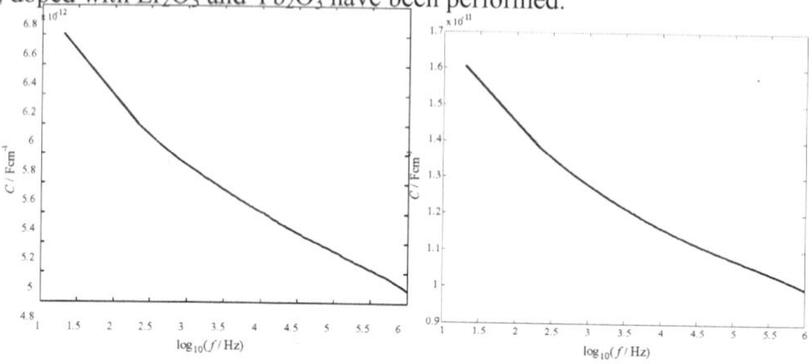

a b

Fig 10 Microcapacitance vs. frequency for $BaTiO_3$ doped with a) 0.5 wt% Er_2O_3 b) 0.5 wt% Yb_2O_3.

In order to calculate equivalent impedance of the sample, conductance as another dominant electrical parameter shuold be taken into account. It is a parasitic parameter that is given in terms of capacitance with $\tan\delta$ as a measure of losses, i.e., $1/R=G=\omega C \cdot \tan\delta$. The intergranular impedance model also contains two additional parameters; inductance L, and capacitance C_p. Their nature cannot be correlated with geometrical parameters of grains in general way.

In order to determine an algebaric equation describing equivalent intergranular impedance in terms of circuit parameters following equation can be used:

$$Z(s) = \frac{1 + CR \cdot s + CL \cdot s^2}{(Cp + C) \cdot s + CpCR \cdot s^2 + CpCL \cdot s^3}$$

Based on proposed equivalent circuit and the theory of impedance analysis for the model of three aggregate spheres the equivalent impedance can be defined by:

$$Z_e = \frac{Z_{12} \cdot (Z_{13} + Z_{23})}{Z_{12} + Z_{13} + Z_{23}}$$

where Z_{12}, Z_{13}, Z_{23} are the intergranular impedances between two adjacent particles. Then, this model can be inserted for any contact region inside the multi-particle model system during its microstructure development. Thus, electrical properties are determined in general by a series combination of such impedances.

CONCLUSIONS

In this study, the model of intergranular impedance is established using the equivalent electrical scheme characterized by corresponding frequency characteristic. According to the microstructures we have obtained for $BaTiO_3$ doped with Er_2O_3 and Yb_2O_3 the global impedance of barium-titanate ceramics sample, which contains both resistor and capacity component, has been presented as a "sum" of many clusters of micro-resistors and micro-capacitors connected in tetrahedral lattice. The positions of neighboring grains for the four grains cluster have been defined and according to them the tetrahedral scheme of mutual electrical influence of $BaTiO_3$ grains has been established. Fractal geometry has been used to describe complexity of the spatial distribution of $BaTiO_3$ grains. The model of impedances between clusters of ceramics grains has been presented and calculations of microcapacitance generated in grains contacts of $BaTiO_3$ doped with Er_2O_3 and Yb_2O_3 have been performed. By the control of shapes and numbers of contact surfaces on the level of the entire $BaTiO_3$–ceramic sample, the control over structural properties of this ceramics can be done, with the aim of correlation between material electronic properties and corresponding microstructure.

ACKNOWLEDGMENTS

This research is a part of the project "Investigation of the relation in triad: synthesis-structure-properties for functional materials" (No.142011G). The authors gratefully acknowledge the financial support of Serbian Ministry for Science for this work.

REFERENCES

[1] M.M.Vijatovic, J.D.Bobic, B.D.Stojanovic History and Challenges of Barium Titanate I Sci.Sint Vol.40 2 (2008) 155-167

[2] C.Pithan, D.Hennings, R. Waser Progress in the Synthesis of Nanocrystalline $BaTiO_3$ Powders for MLCC International Journal of Applied Ceramic Technology 2 (1), (2005), 1–14

[3] V.V. Mitić, I. Mitrović, D. Mančić, "The Effect of $CaZrO_3$ Additive on Properties of $BaTiO_3$-Ceramics", Sci. Sint., Vol. 32 (3), pp. 141-147, 2000

[4] P.W.Rehrig, S.Park, S.Trolier-McKinstry, G.L.Messing, B.Jones, T.Shrout Piezoelectric properties of zirconium-doped barium titanate single crystals grown by templated grain growth J. Appl. Phys. Vol 86 3, (1999) 1657-1661

[5] S. Wang, G.O. Dayton Dielectric Properties of Fine-Grained Barium Titanate Based X7R Materials J. Am. Ceram. Soc. 82 (10), (1999), 2677–2682

[6] V.P.Pavlovic, M.V.Nikolic, V.B.Pavlovic, N. Labus, Lj. Zivkovi}, B.D.Stojanovic, Correlation between densification rate and microstructure evolution of mechanically activated $BaTiO_3$, Ferroelectrics 319 (2005) 75-85

[7] D. Lu, X. Sun, M. Toda Electron Spin Resonance Investigations and Compensation Mechanism of Europium-Doped Barium Titanate Ceramics Japanese Journal of Applied Physics Vol. 45, No. 11, 2006, pp. 8782-8788

[8] V.V.Mitić, Lj. M. Kocić, M. Miljković and I. Petković, Fractals and $BaTiO_3$ – Ceramics microstructure analysis, Mikrochim. Acta [Suppl.] **15**, (1998), 365-369

[9] V.Mitic Lj.Kocic, I.Mitrovic, M.M.Ristic Models of $BaTiO_3$ Ceramics Grains Contact Surfaces The 4th IUMRS International Conference in Asia OVTA Makuhari, Chiba, Japan !997

[10] B. Mandelbrot, *The Fractal Geometry of Nature*, W. H. Freeman and Co., New York, 1983.

DIELECTRIC PROPERTIES OF $ACu_3Ti_4O_{12}$-TYPE PEROVSKITES

Matthew C. Ferrarelli, Derek C. Sinclair and Anthony R. West.
Department of Engineering Materials, University of Sheffield,
Mappin Street, Sheffield, United Kingdom, S1 3JD.

ABSTRACT

The synthesis, ceramic microstructure and electrical properties of $ACu_3Ti_4O_{12}$-type perovskites where A = Ca (CCTO), $Na_{1/2}Bi_{1/2}$ (NBCTO) and $Bi_{2/3}$ (BCTO) are presented. Phase-pure samples of CCTO and NBCTO were prepared from nominal stoichiometric compositions, however, a non-stoichiometric starting composition with 6 mol % TiO_2 deficiency was required to obtain phase-pure BCTO (BCTO'). All samples exhibit the so-called 'giant permittivity' effect at room temperature with ε' > 4000 at radio-frequencies. Impedance Spectroscopy (IS) reveals all ceramics to be electrically heterogeneous, consisting of semiconducting grains and insulating grain boundaries. The giant permittivity effect is attributed to an internal or grain boundary barrier layer capacitance mechanism.

The presence of the stereochemically active Bi^{3+} ion on the A-site has a dramatic effect on the magnitude of the intrinsic, or bulk relative permittivity (ε_r) for $ACu_3Ti_4O_{12}$-type perovskites. ε_r ~ 100 for CCTO but exceeds 200 for NBCTO and BCTO'. ε_r for all compounds is significantly higher than that calculated using the Clausius-Mossotti equation, implying the presence of a polarisation mechanism in addition to the polarisability of the constituent ions. This mechanism is attributed to incipient ferroelectricity.

INTRODUCTION

The crystal structure of $ACu_3Ti_4O_{12}$ compounds[1-3] where A = Ca^{2+}, $Na_{1/2}Bi_{1/2}$, $Na_{1/2}La_{1/2}$, $Bi_{2/3}$, $(RareEarth)_{2/3}$ etc. can be described as consisting of a tilted three dimensional corner sharing network of TiO_6 octahedra, with 1:3 cation ordering of the A-site and Cu^{2+} cations, respectively. Tilting of the TiO_6 network is extreme, with a Ti-O-Ti bond angle of 141 °, causing two unequal coordination environments for the A-site and the Cu^{2+} cation. The d^9 Cu^{2+} Jahn-Teller cation is supported in a rigid square planar environment at the face and edge centres of the unit cell, whereas the larger A-site cation occupies a body-centred arrangement in an icosahedral coordination environment, Fig. 1. All of the $ACu_3Ti_4O_{12}$ family crystallise in the cubic centrosymmetric space group, $Im\bar{3}$.

$CaCu_3Ti_4O_{12}$ (CCTO) has recently attracted a lot of attention[4-18] due to the high apparent permittivity (ε' ~ 10,000) exhibited in polycrystalline samples. Impedance Spectroscopy (IS) has shown CCTO ceramics to be electrically heterogeneous, consisting of semiconducting grains and insulating grain boundaries[7,12-14]. The high apparent permittivity can therefore be explained by an internal barrier layer capacitance (IBLC) mechanism, where potential barriers are formed between semiconducting grains and insulating grain boundaries. Initial permittivity measurements by Subramanian *et al*[1] on a range of $ACu_3Ti_4O_{12}$ compounds established CCTO as having the highest room temperature effective permittivity at a frequency of 100 kHz. The majority of research into CCTO has focused on the characterisation and optimisation of the extrinsic grain boundary barrier layer mechanism and minimal work has focused on the intrinsic bulk properties of $ACu_3Ti_4O_{12}$-type perovskites. In this study, both the extrinsic IBLC effect and intrinsic bulk properties of CCTO and two A-site analogues, $Na_{0.5}Bi_{0.5}Cu_3Ti_4O_{12}$ (NBCTO) and '$Bi_{0.67}Cu_3Ti_4O_{12}$' (BCTO), are investigated.

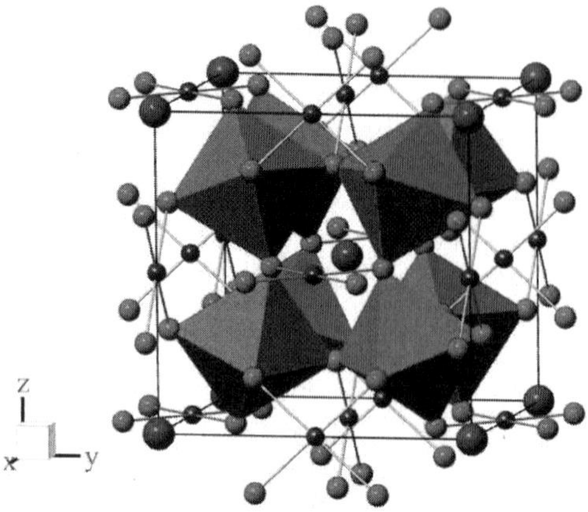

Figure 1. Crystal structure of $ACu_3Ti_4O_{12}$. A = large (blue) spheres, Cu = small (black) spheres (note the square planar bonding), O = small (red) spheres, TiO_6 = green octahedra.

EXPERIMENTAL

Powders of the three target compounds and $Bi_{2/3}Cu_3Ti_{3.84}O_{11.68}$ (BCTO') were prepared by the conventional mixed oxide route. High-purity grade Na_2CO_3, Bi_2O_3, $CaCO_3$, CuO and TiO_2 powders (all > 99 % purity, Aldrich Chemical Co., Milwaukee, WI) were dried prior to use and then weighed in appropriate quantities to make ~ 3 – 5 g batches of the desired compounds. The batches were mixed with acetone in an agate mortar and pestle and then heated at high temperature in a muffle furnace. Reaction temperatures and periods required were as follows: 1000 °C for 48 h (CCTO); 600 °C for 24 h followed by 975 °C for 36 h (NBCTO); 600 °C for 24 h followed by 950 °C for 36 h (BCTO and BCTO'). Pellets of the resulting powder for CCTO, NBCTO and BCTO' were uniaxially cold pressed into 5 mm diameter compacts and sintered (all for 6 h) at 1100, 1050 and 1000 °C, respectively with a heating and cooling rate of 300 $°Ch^{-1}$. Phase purity and crystallinity of the powders and the ceramics was determined by X-Ray Diffraction (XRD) using a high-resolution diffractometer ($CuK\alpha_1$, 1.54059 Å, Model Stoe StadiP, Stoe and Cie GmbH, Darmstadt, Germany) operated at 50 kV and 30 mV (step size of 0.02 ° and scan rate of 2 $°min^{-1}$). The ceramic microstructure of as-fired surfaces was examined using a scanning electron microscope (SEM) (JEOL 6400, JEOL Corporation, Tokyo, Japan) operated at 15 kV.

Impedance Spectroscopy and fixed frequency capacitance measurements were performed over the range ~ 10 – 600 K using an impedance analyser (Hewlett Packard 4192A, Hewlett-Packard Co., Palo Alto, CA)) and an LCR bridge (Model HP4284, Hewlett-Packard, Palo Alto, CA), respectively. An AC voltage amplitude of 100 mV was used and the major faces of the sintered ceramics were polished prior to applying electrodes. Gold sputtered electrodes were applied using an Emscope SC500 gold sputter-coater and were deposited on each side of the ceramic for 8 min at a current of 20 mA under an argon atmosphere. For high temperature measurements, samples were supported in a tube furnace by a compression jig and were connected to the jig using platinum foil pressed onto the parallel faces of the pellet. To obtain low temperature measurements, an Oxford Instruments CCC1104 cryostat built on an Edwards model CS2/9 cold head was used, with an Edwards Helium Cryodrive 1.5 compressor used for cooling. The samples were placed on the cryostat stage and attached to the instrument via gold connectors. Data were corrected for the geometric factor of the sample (sample thickness / electrode area), and additionally, high temperature measurements were corrected for the stray electrical contributions of the measuring apparatus. All data were analysed using the commercial software package Z-view (Scribner Associates, Inc., Charlottesville, VA, Version 2.1).

RESULTS

XRD patterns of CCTO, NBCTO and BCTO' sintered pellets (not shown) were fully indexed in the space group $Im\bar{3}$, with a = 7.3974(2), 7.4114(3) and 7.4205(1) Å, respectively. BCTO could not be prepared as a single-phase powder or sintered ceramic. In both cases, the XRD pattern showed the presence of TiO_2 (rutile) as a secondary phase. SEM micrographs of the ceramic microstructures are shown in Fig. 2. CCTO and NBCTO have similar microstructures with an average grain size in the range 2 – 10 μm. BCTO' has a smaller average grain of ~ 1 – 5 μm. Pellet density was ~ 90 % of the theoretical x-ray density for all samples.

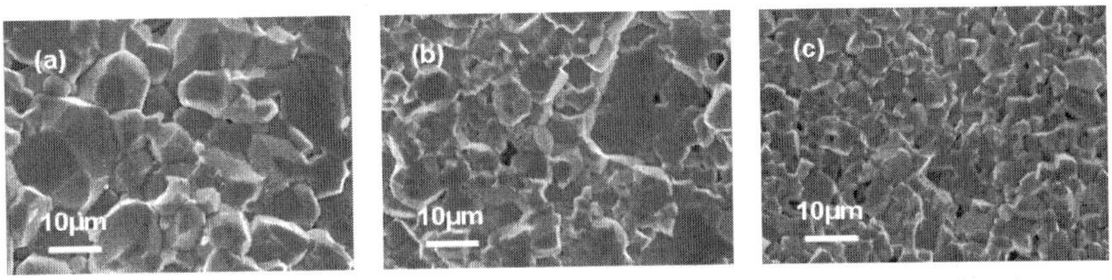

Figure 2. SEM micrographs of (a) CCTO, (b) NBCTO and (c) BCTO' ceramics.

The real component of the permittivity, ε', versus temperature for CCTO, NBCTO and BCTO' ceramics at a fixed frequency of 100 kHz is shown in Fig. 3. All ceramics show the so-called 'giant permittivity' effect with ε' > 4000 at 320 K. The effect is smaller for BCTO' compared to CCTO and NBCTO. The inset in Fig. 3 shows the low temperature ε' data at a fixed frequency of 1 MHz on an expanded scale and reveals two interesting features. Firstly, ε' is significantly higher (> 200) for NBCTO and BCTO' compared to CCTO (~ 100). Secondly, ε' decreases with increasing temperature in this range for the Bi-based compounds.

IS data for all three samples could be modelled on an equivalent circuit consisting of two parallel Resistor-Capacitor (RC) elements connected in series. Ideally, for each RC element a semicircular arc results in the Z* plot. One RC element represents the bulk (intra-grain) response (R_bC_b), where C_b ~ 10 – 30 $pFcm^{-1}$ and the other represents the grain boundary response ($R_{gb}C_{gb}$), where C_{gb} ~ 0.3 – 0.9 $nFcm^{-1}$. Representative data are shown in Fig. 4 for BCTO' at 25 and 523 K. The arc observed in the Z* plot at low temperature, Fig. 4 (a), and the non-zero intercept on the real axis of the Z* plot at high temperature, inset in Fig. 4 (b), represent the bulk response at low and high temperature, respectively. R_b ~ 400 kΩcm and C_b ~ 30 $pFcm^{-1}$ at 25 K and R_b ~ 90 Ωcm at 523 K for BCTO'. The arc in the Z* plot in Fig. 4 (b) represents the grain boundary response for BCTO' with R_{gb} ~ 25 kΩ cm and C_{gb} ~ 0.3 $nFcm^{-1}$, at 523 K. It is noteworthy that C_{gb} decreases in the order NBCTO (~ 0.94 $nFcm^{-1}$, ε' ~ 10,600) > CCTO (~ 0.84

nFcm^{-1}, ε' ~ 9,500) > BCTO' (~ 0.35 nFcm^{-1}, ε' ~ 4,000). C_b is significantly larger for NBCTO and BCTO' (~ 20 – 30 pFcm^{-1}, ε_r ~ 225 – 340) compared to CCTO (~ 10 pFcm^{-1}, ε_r ~ 110).

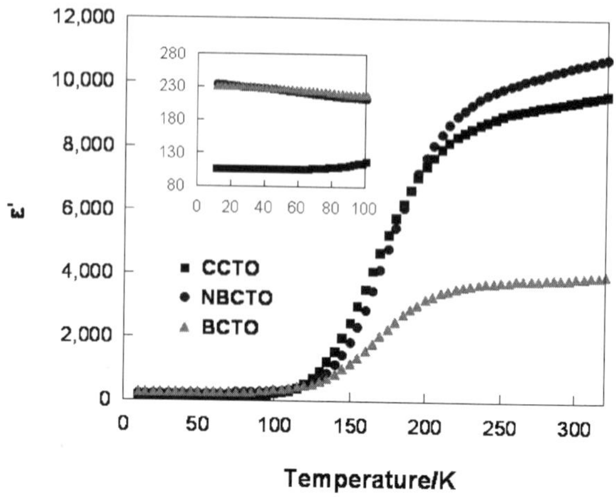

Figure 3. ε' versus temperature at 100 kHz for CCTO, NBCTO and BCTO'. Inset shows ε' at 1 MHz on an expanded scale for 10 – 100 K.

The temperature dependence of the bulk ($\sigma_b = 1/R_b$) and grain boundary ($\sigma_{gb} = 1/R_{gb}$) conductivity for all samples is plotted in Arrhenius format, Fig. 5. σ_{gb}'s obey the Arrhenius law with activation energies, E_a, of ~ 0.67(2) eV (CCTO), 0.52(2) eV (NBCTO) and ~ 0.87(3) eV (BCTO'). σ_b's show significant deviation from the Arrhenius law, especially for NBCTO and BCTO'. E_a ~ 0.1 eV for the higher temperature data, however, the bulk conduction process appears to have very little thermal dependence at low temperatures for NBCTO and BCTO'.

The temperature dependence of C_b below ~ 140 K was investigated by plotting the IS data in the form of spectroscopic plots of the imaginary component of the electric modulus, M''. For an ideal, parallel R_bC_b element, a Debye peak occurs in the M'' spectrum and $C_b = (2 \cdot M''_{max})^{-1}$. M''_{max} increases in height with increasing temperature for all samples, Fig. 6 and therefore C_b and ε_r decrease with increasing temperature, Fig. 7. It is noteworthy that ε_r is significantly higher for the Bi-based compounds.

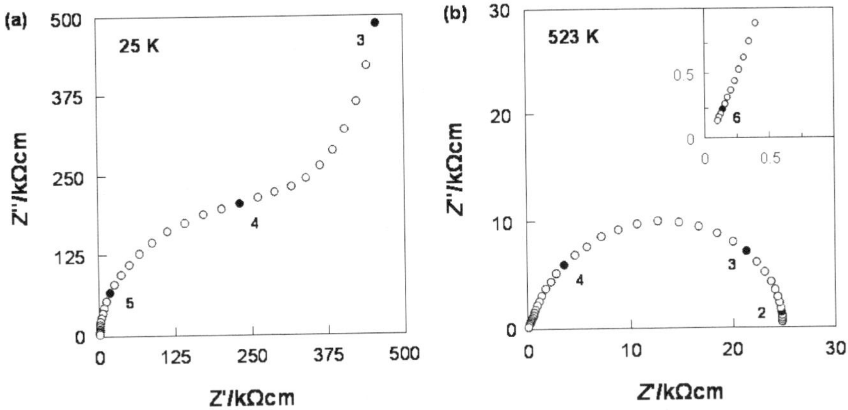

Figure 4. Z* plots for BCTO' ceramics at (a) 25 K and (b) 523 K. Selected frequencies on a log (f/Hz) scale are shown by filled symbols.

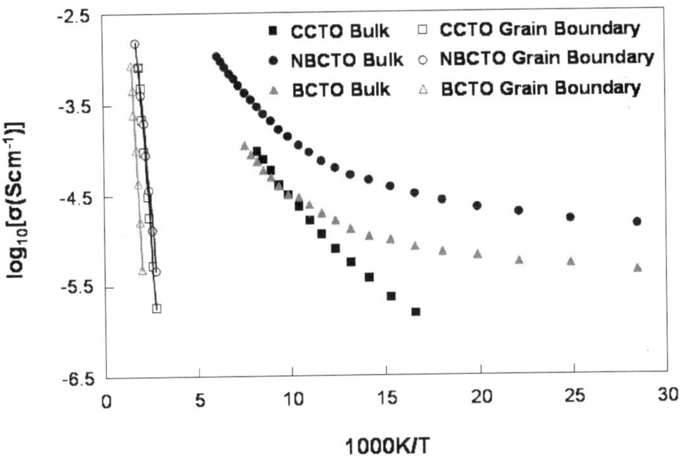

Figure 5. Arrhenius plots of bulk and grain boundary conductivity.

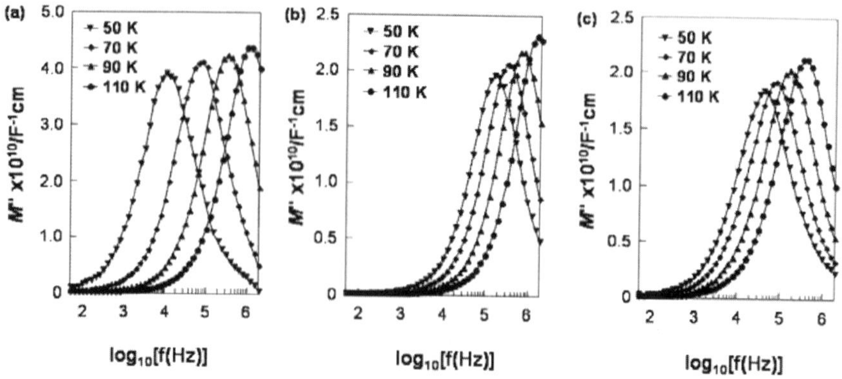

Figure 6. M'' spectroscopic plots at selected temperatures for (a) CCTO, (b) NBCTO and (c) BCTO' ceramics.

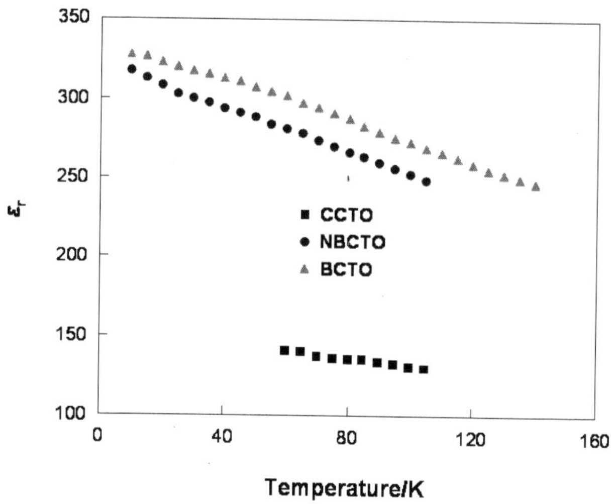

Figure 7. Temperature dependence of ε_r for CCTO, NBCTO and BCTO' ceramics. Values obtained from M'' spectroscopic plots.

DISCUSSION

The lattice parameter obtained for CCTO, a = 7.3974(2) Å, is slightly larger than that reported by Subramanian *et al*[1], but is within the range of ~ 7.3798(1) – 7.405 Å reported by other researchers[10]. NBCTO has a = 7.4114(3) Å, a value almost identical to that reported by Subramanian and Sleight[3], a = 7.412 Å, but the lattice parameter for BCTO', a = 7.4205(1) Å, is larger than that reported[1] for BCTO, a = 7.413 Å. The composition $Bi_{2/3}Cu_3Ti_4O_{12}$ was first synthesized by Bryntse and Werner[11], with a reported lattice parameter of 7.4175(3) Å, however, they noted that the product was not phase-pure and contained a small impurity of TiO_2. In subsequent work[1,9], there is no mention of a TiO_2 impurity. Attempts at preparing phase-pure, stoichiometric $Bi_{2/3}Cu_3Ti_4O_{12}$ in our laboratory were unsuccessful, the products always contained TiO_2 as a secondary phase. Reducing the amount of TiO_2 in the starting composition ($Bi_{2/3}Cu_3Ti_{4-x}O_{12-2x}$) was always required to obtain single-phase samples, e.g. $Bi_{2/3}Cu_3Ti_{3.76}O_{11.52}$ (x = 0.24). Although not reported in detail here, single-phase samples have been prepared in the range $0.1 \leq x \leq 0.4$. The presence of TiO_2 as an impurity phase has also been reported for other $A_{2/3}Cu_3Ti_4O_{12}$ compositions[15], where A = La to Ho. These findings suggest that significant non-stoichiometry may exist in the so-called '$A_{2/3}Cu_3Ti_4O_{12}$'-type compounds and this merits further investigation.

All ceramics are electrically heterogeneous and consist of semiconducting grains and insulating grain boundaries, see Figs. 4 and 5. Such an electrical microstructure gives rise to the IBLC mechanism that is responsible for the giant permittivity effect shown in Fig. 3. The variation in the magnitude of ε' near room temperature can be explained by changes in the microstructure. BCTO' has a smaller average grain size compared to CCTO and NBCTO, Fig. 2, and for an IBLC mechanism this results in a lower value of ε'. As grain boundary capacitances, C_{gb}, in barrier layer capacitors are dependent on the ceramic microstructure, results in the literature show considerable variation depending on the processing conditions employed. For example, C_{gb} of CCTO ceramics range from 0.43 $nFcm^{-1}$ (ε' ~ 4,900)[12] for fine-grained ceramics to 25 $nFcm^{-1}$ (ε' ~ 280,000)[13] for coarse grained ceramics. The NBCTO and BCTO' ceramics in this study exhibit larger ε' values at room temperature compared to that reported elsewhere, i.e. 2,454[3] and 1,871[1], respectively. The simplest explanation for this 'discrepancy' is a variation in ceramic microstructures due to different sintering conditions being employed in the different studies. No attempt has been made here to maximise the giant permittivity effect in these compounds, however, an IBLC mechanism appears to be present in all undoped $ACu_3Ti_4O_{12}$ and $A_{2/3}Cu_3Ti_4O_{12}$-based ceramics reported to date[4,5,7,9,12-14,16].

The activation energies for bulk conduction near room temperature, ~ 0.1 eV and the grain boundary E_a of ~ 0.5 – 0.8 eV, Fig. 5, are similar to those reported in the literature for CCTO-type materials[7,9,12-14,16]. In addition, the non-Arrhenius-type behaviour of σ_b at low temperatures has also been reported in other studies[16]. The origin of the semiconductivity in CCTO and its derivatives remains unsolved as does the composition of the grain boundary regions. A detailed discussion on the defect chemistry of these materials is clearly outside the scope of the present manuscript, however, we note that the Ti-site non-stoichiometry associated with BCTO' does not have a significant influence on the occurrence of the bulk conduction mechanism, Fig. 5. Further work is still required to understand the origin and mechanism(s) of the bulk conductivity in CCTO-type materials.

There is a significant difference in ε_r of CCTO compared to NBCTO and BCTO'. ε_r is approximately twice as large for the Bi-based compounds compared to CCTO, see inset in Fig. 3

and Figs. 6 and 7. Furthermore, all ε_r values are much higher than those expected from calculations using the Clausius-Mossotti equation where $\varepsilon_r \sim 48$ (CCTO), 58 (NBCTO) and 44 (BCTO'). ε_r is therefore at least a factor of four higher than expected for the Bi-based compounds and is approximately double that expected for CCTO. This indicates an additional polarisation mechanism other than simple ion polarisabilities must be present in these compounds.

The increase in ε_r for NBCTO and BCTO' compared to CCTO can be partially explained by the replacement of Ca^{2+} ions on the A-site of the lattice with the more polarisable Bi^{3+} ions but this does not explain the difference adequately. Incorporation of Bi^{3+} ions on the A-site of many perovskite-based materials is known to increase their permittivity by the off-centre movement(s) of the stereochemically active Bi^{3+} ion within its coordination sphere, creating a localised dipole moment. This movement is caused by the interaction of the lone electron pair on the Bi^{3+} ion with the surrounding O^{2-} anions. This A-site mechanism can therefore be used to increase the intrinsic permittivity in $ACu_3Ti_4O_{12}$-type compounds.

Recent results on Mn-doped CCTO ceramics[18] have revealed evidence for incipient ferroelectricity in CCTO. Although the origin of the incipient ferroelectricity remains unclear, this explains the higher than expected and temperature dependent behaviour of ε_r for undoped CCTO, Fig. 7. The ε_r data for NBCTO and BCTO are also temperature dependent which suggests incipient ferroelectricity also occurs in these compounds. Although these compounds are all cubic with centrosymmetric symmetry, they are all based on a crystal structure consisting of a tilted, TiO_6, octahedral network, Fig. 1. Many related perovskites that consist of tilted TiO_6 octahedra, such as $CaTiO_3$, are known to exhibit incipient ferroelectricity. We propose that higher than expected ε_r values for $ACu_3Ti_4O_{12}$-type perovskites are associated with incipient ferroelectricity and that this is an 'intrinsic' feature associated with the unusual perovskite-type crystal structure of these compounds.

CONCLUSIONS

Electrical measurements have shown two A-site analogues of CCTO, NBCTO and BCTO', to exhibit the same giant permittivity effect at room temperature as CCTO. Ceramics of all compounds are electrically heterogeneous, consisting of semiconducting grains and insulating grain boundaries. The origin of the giant permittivity effect is therefore attributed to an Internal Barrier Layer Capacitance mechanism associated with the electrically heterogeneous nature of the ceramics. The grain boundary capacitance of NBCTO and BCTO' ceramics is higher than that reported previously[1,3], with NBCTO displaying a grain boundary capacitance higher than that of CCTO. These variations are attributed to differences in ceramic microstructure. The bulk capacitance of CCTO, NBCTO and BCTO ceramics show values higher than that expected from the Clausius-Mossotti equation. Additionally, the bulk capacitances of NBCTO and BCTO are double that of CCTO. All $ACu_3Ti_4O_{12}$-type compounds are proposed to exhibit incipient ferroelectricity and the higher capacitance of the Bi-based compounds is attributed to stereochemical activity associated with off-centre movements of the Bi^{3+} ions on the A-site sublattice.

ACKNOWLEDGEMENTS

We thank the EPSRC and the EU (NUOTO-project) for funding.

REFERENCES

[1]M.A. Subramanian, D. Li, N. Duan, B.A. Reisner, and A.W. Sleight, High Dielectric Constant in $ACu_3Ti_4O_{12}$ and $ACu_3Ti_3FeO_{12}$ Phases, *J. Solid State Chem.*, **151**, 323-5 (2000).
[2]A. Deschanvres, B. Raveau, and F. Tollemer, Substitution of Copper for a Bivalent Metal in Titanates of Perovskite Type, *Bull. Soc. Chim. Fr.*, 4077-81 (1967).
[3]M.A. Subramanian, and A.W. Sleight, $ACu_3Ti_4O_{12}$ and $ACu_3Ru_4O_{12}$ Perovskites: High Dielectric Constants and Valence Degeneracy, *Solid State Sci.*, **4**, 347-51 (2002).
[4]R. Zuo, L. Feng, Y. Yan, B. Chen, and G. Cao, Observation of Giant Dielectric Constant in $CdCu_3Ti_4O_{12}$ Ceramics, *Solid State Commun.*, **138**, 91-4 (2006).
[5]A.P. Ramirez, M.A. Subramanian, M. Gardel, G. Blumberg, D. Li, T. Vogt, and S.M. Shapiro, Giant Dielectric Constant Response in a Copper-Titanate, *Solid State Commun.*, **115**, 217-20 (2000).
[6]C.C. Homes, T. Vogt, S.M. Shapiro, S. Wakimoto, and A.P. Ramirez, Optical Response of High-Dielectric-Constant Perovskite-Related Oxide, *Science*, **293**, 673-76 (2001).
[7]D.C. Sinclair, T.B. Adams, F.D. Morrison, and A.R. West, $CaCu_3Ti_4O_{12}$: One-Step Internal Barrier Layer Capacitor, *Appl. Phys. Lett.*, **80**, 2153-5 (2002).
[8]S.-Y. Chung, I.-D. Kim, and S.-J.L. Kang, Strong Nonlinear Current-Voltage Behaviour in Perovskite-Derivative Calcium Copper Titanate, *Nature Mater.*, **3**, 774-8 (2004).
[9]J. Liu, C.-G. Duan, W.-G. Yin, W.N. Mei, R.W. Smith, and J.R. Hardy, Large Dielectric Constant and Maxwell-Wagner Relaxation in $Bi_{2/3}Cu_3Ti_4O_{12}$, *Phys. Rev. B*, **70**, 144106–1-7 (2004).
[10]L. Wu, Y. Zhu, S. Park, S. Shapiro, and G. Shirane, Defect Structure of the High-Dielectric-Constant Perovskite $CaCu_3Ti_4O_{12}$, *Phys. Rev. B*, **71**, 014118–1-7 (2005).
[11]I. Bryntse, and P.-E. Werner, Synthesis and Structure of a Perovskite Related Oxide, $Bi_{2/3}Cu_3Ti_4O_{12}$, *Mat. Res. Bull.*, **25**, 477-83 (1990).
[12]R.K. Grubbs, E.L. Venturini, P.G. Clem, J.J. Richardson, B.A. Tuttle, and G.A. Samara, Dielectric and Magnetic Properties of Fe-and Nb-doped $CaCu_3Ti_4O_{12}$, *Phys. Rev. B*, **72**, 104111–1-11 (2005).
[13]T.B. Adams, D.C. Sinclair, and A.R. West, Giant Barrier Layer Capacitance Effects in $CaCu_3Ti_4O_{12}$ Ceramics, *Adv. Mater.*, **14**, 1321-3 (2002).
[14]T.B. Adams, D.C. Sinclair, and A.R. West, Characterization of Grain Boundary Impedances in Fine- and Coarse-Grained $CaCu_3Ti_4O_{12}$ Ceramics, *Phys. Rev. B*, **73**, 094124–1-9 (2006).
[15] B. Bochu, M.N. Deschizeaux, J.C. Joubert, A. Collomb, J. Chenavas, and M. Marezio, Synthèse et Caractérisation d'une Série de Titanates Pérowskites Isotypes de $[CaCu_3](Mn_4)O_{12}$, *J. Solid State Chem.*, **29**, 291-8 (1979).
[16]G. Chiodelli, V. Massarotti, D. Capsoni, M. Bini, C.B. Azzoni, M.C. Mozzati, and P. Lupotto, Electric and Dielectric Properties of Pure and Doped $CaCu_3Ti_4O_{12}$ Perovskite Materials, *Solid State Commun.*, **132**, 241-6 (2004).
[17]M. Li, A. Feteira, D.C. Sinclair, and A.R. West, Influence of Mn-Doping on the Semiconducting Properties of $CaCu_3Ti_4O_{12}$ Ceramics, *Appl. Phys. Lett.*, **88**, 232903–1-3 (2006).
[18]M. Li, A. Feteira, D.C. Sinclair, and A.R. West, Incipient Ferroelectricity and Microwave Dielectric Properties of $CaCu_{2.85}Mn_{0.15}Ti_4O_{12}$ Ceramics, *Appl. Phys. Lett.*, **91**, 132911–1-3 (2007).

DIELECTRIC PROPERTIES OF RARE EARTH DOPED Sr-M HEXAFERRITES

Anterpreet Singh[1], S. Bindra Narang[2], Kulwant Singh[1] and [3]R.K. Kotnala

[1]Department of Physics, Guru Nanak Dev University, Amritsar, India
[2]Department of Electronics and Technology, Guru Nanak Dev University, Amritsar, India

[3] Magnetic Materials & Standards, National Physical Laboratory, New Delhi, India

ABSTRACT

Strontium hexaferrites of a structural formula $Sr_{1-x}RE_xFe_{12}O_{19}$, where RE = La^{3+}, Nd^{3+} and Sm^{3+} with (x = 0 to 0.30) were prepared by a standard ceramic technique. The lattice constants, density, porosity, dielectric constant and dielectric loss tangent were studied on a series of rare earth substituted strontium hexaferrites. The dielectric constant and dielectric loss tangent were measured in the frequency range of 20 Hz to 1 MHz at room temperature. The dielectric constant decreased with increasing frequency for all the three series. This behavior of dielectric properties with frequency has been explained with the Maxwell–Wagner type interfacial polarization in agreement with the Koops phenomenological theory. The substitution of rare earth ions into $SrFe_{12}O_{19}$ increases the value of the dielectric constant. This increase in dielectric constant could be due to the electronic exchange between $Fe^{2+} \leftrightarrow Fe^{3+}$ and results in a local displacement determining the polarization of the ferrites. For these ferrite samples the Curie temperature decreases as rare earth ions substitution increases.

INTRODUCTION

The hexagonal ferrites, $MFe_{12}O_{19}$ (M = Ba, Sr, Pb) have been extensively studied during the past years, because of their technological interest as traditional permanent magnets, microwave device materials and magnetic recording media [1,2]. From the crystal chemistry point of view the ferrites have an interesting and complex-magnetoplumbite type structure in which the iron ions are coordinated tetrahedrally (FeO_4), trigonal bipyramidally (Fe_2O_5) and octahedrally (FeO_6) by oxygen ions. For years, many researchers have studied the magnetic properties of each Ba-ferrite and Sr-ferrite prepared by various techniques such as chemical coprecipitation method [3], the glass crystallization [4], the salt melt method [5], the sol-gel method [6] and ceramic process [7]. Most of the research has emphasized the modification of magnetic properties by the substitution of Fe^{3+} with 3d ions such as Co^{2+} Ti^{4+}, Co^{2+} Sn^{4+} and Cr^{3+} Al^{3+} [8, 9]. Magnetic properties of RE substitution in the $SrFe_{12}O_{19}$ system have been reported [10, 11]. However, investigations of substitutional effect on the dielectric properties of strontium ferrite are rare. In the present investigation, focus has been made to study the effect of RE^{3+} substitution on the dielectric properties of strontium ferrite to understand the conduction mechanism of these ferrites. RE^{3+} may substitute Sr^{2+} at crystallographic site due to the smaller ionic radii of RE^{3+} with Sr^{2+}. The substitution of divalent Sr^{2+} ion by trivalent RE^{3+} ion will change Fe^{3+} ion to Fe^{2+} ion per formula unit, which may enhance its dielectric properties also.

EXPERIMENTAL

A series of $Sr_{1-x}RE_xFe_{12}O_{19}$ samples with different substitution ratios were prepared by a standard ceramic processing technique. High purity precursors $SrCO_3$, RE_2O_3 and Fe_2O_3 were mixed together in the appropriate molar ratio, calculated from the following chemical reaction

$$(1\text{-}x)\ SrCO_3 + 6\ Fe_2O_3 + (x/2)\ RE_2O_3 \longrightarrow Sr_{1-x}RE_xFe_{12}O_{19} + (1\text{-}x)\ CO_2 + (x/4)\ O_2$$

where, x varies from 0 to 0.30 with an increment of 0.10. The details of the preparation method have been given in our earlier publication [12]. The phase analysis of these sintered pellets was carried out by X-ray diffractometer. Dielectric measurements of all ferrite samples were done in the frequency range from 20 Hz- 1 MHz using precision LCR meter model (HP4284A) by standard two-probe technique using platinum electrodes.
The value of dielectric constant (ε′) of the ferrite samples can be calculated by using the formula

$$\varepsilon' = \frac{C_p t}{\varepsilon_o A} \tag{1}$$

where C_p is the capacitance of samples in pF, t the thickness of the samples in cm, A is the cross-sectional area of the sample in cm^2 and ε_o is the permittivity in free space having value 8.854×10^{-2} pF/cm. Curie temperatures T_c (K) of all the samples were determined by the gravity method [13].

RESULTS AND DISCUSSION

X-ray diffraction

Figure 1 shows a sequence of X-ray diffraction patterns obtained at different molar concentrations of $Sr_{1-x}RE_xFe_{12}O_{19}$ (RE = La^{3+}, Nd^{3+} and Sm^{3+}). The analysis reveals that in case of La^{3+} substitution, the samples are single hexagonal M-type phase (Fig. 1a). No peaks of Fe_2O_3 and La_2O_3 phases are observed, which suggest that Sr^{2+} ion is substituted by La^{3+} ions [14]. In case of Nd^{3+} series, all peaks correspond to hexagonal M-type phase. However, for the substitution x = 0.30, a weak peak characteristic of the hematite (α-Fe_2O_3) phase is observed, indicating that the sample contains a slight proportion of α-Fe_2O_3 (Fig. 1b). In case of Sm^{3+} series, all peaks correspond to hexaferrites, however for the substitution x = 0.30, extra peaks of hematite (α-Fe_2O_3) and tetragonal $Sr_3Fe_2O_7$ are observed (Fig. 1c). This indicates that Nd and Sm

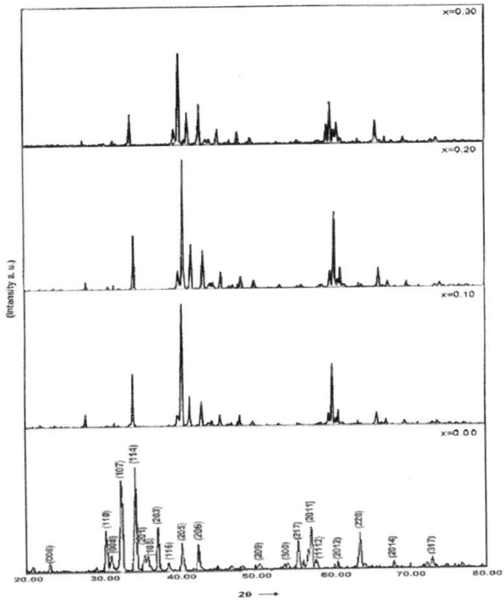

(a) RE=La^{3+}

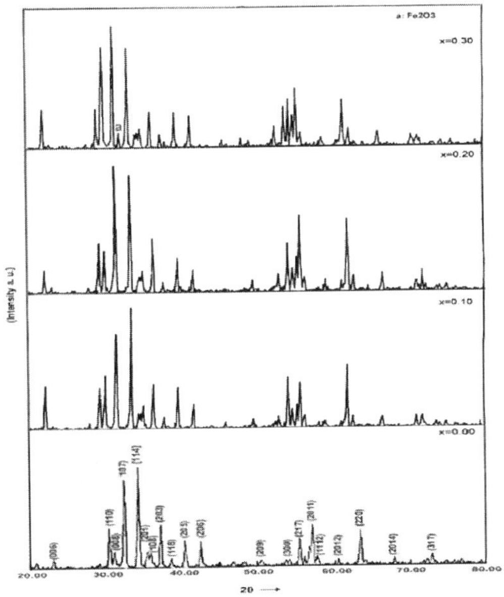

(b) RE=Nd^{3+}

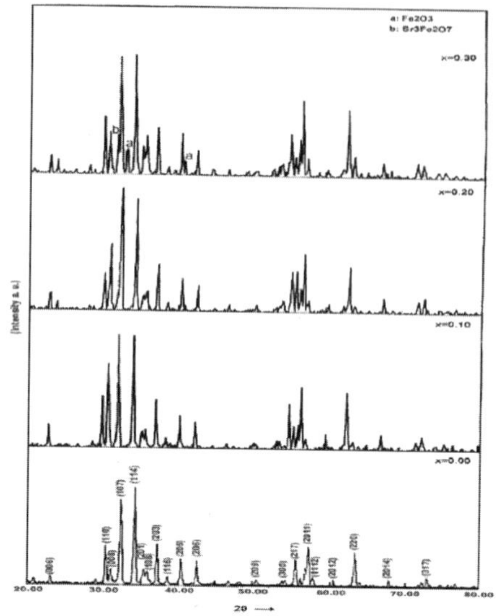

(c) RE=Sm^{3+}

Figure1. X-ray diffraction pattern for the $Sr_{1-x}RE_xFe_{12}O_{19}$ where RE = La^{3+}, Nd^{3+} and Sm^{3+} with (x = 0 to 0.30) for La^{3+} (a), Nd^{3+} (b) and Sm^{3+} (c) respectively.

for x = 0.30 did not substitute totally into the Sr M-type structure resulting in incomplete reactions between Fe^{3+} and Sr^{2+}, indicated by tracing of secondary phases in these samples, and is attributed to the preparation process.

Lattice Constants

X-ray diffraction patterns of $Sr_{1-x}RE_xFe_{12}O_{19}$ hexagonal ferrites of three series under investigation have been obtained using Cu-Kα radiation.
The lattice constants 'a' and 'c' were calculated by the following equation

$$d_{(hkl)} = \left(\frac{4}{3} \frac{h^2 + hk + k^2}{a^2} + \frac{l^2}{c^2} \right)^{-\frac{1}{2}} \qquad (2)$$

The variation of lattice constant 'a' and 'c' with composition (x) for three series prepared with RE= La^{3+}, Nd^{3+} and Sm^{3+} are shown in the figures 2 and 3 respectively. It was observed that both 'a' and 'c' decrease continuously with increasing substituted amount of rare earth ions for the three series. The observed variation in the lattice constants can be explained on the basis of relative ionic radii of Sr^{2+} ions and RE^{3+} ions, which are (1.27 Å) for Sr^{2+} and (1.22 Å, 1.16 Å and 1.13 Å) for La^{3+}, Nd^{3+} and Sm^{3+} respectively [15]. Since RE^{3+} ions have ionic radii less that of the ionic radii of Sr^{2+} ions, the replacement of Sr^{2+} ions by RE^{3+} ions results in the decrease of unit cell dimensions of hexagonal lattice.

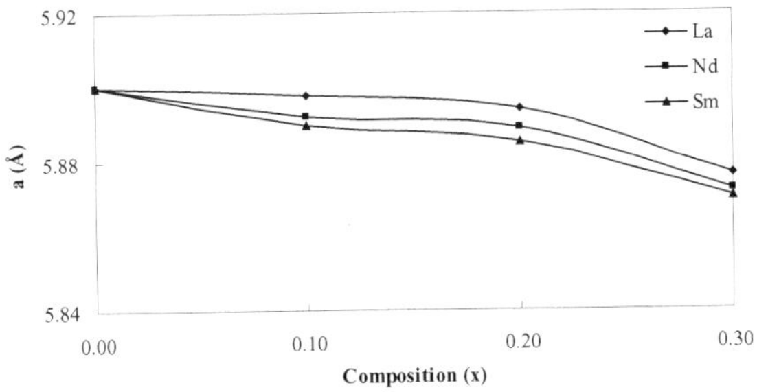

Figure 2. Variation of lattice constant 'a' with composition (x) for three series.

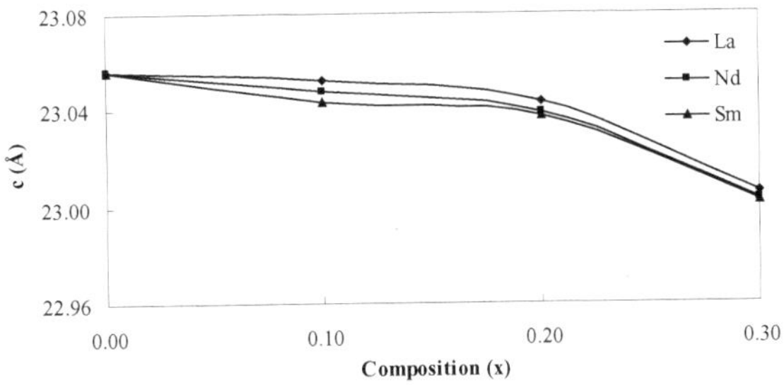

Figure 3. Variation of lattice constant 'c' with composition (x) for three series.

The more decrease in lattice constants 'a' and 'c' for Sm^{3+} ions substituted hexaferrites as compared to those for Nd^{3+} ions and La^{3+} ions substituted ones is attributed to the smaller ionic radii of Sm^{3+} ions than those of Nd^{3+} ions and La^{3+} ions.

Density and Porosity

The X-ray density D_x was calculated by using the known formula

$$D_x = \frac{2nM}{\sqrt{3}N_a a^2 c} \qquad (3)$$

Here n is the number of molecules per unit cell, N_a is the Avogadro's number per gram mole, 'a' and 'c' are the lattice constants obtained from X-ray diffraction analysis and M is the molecular weight of the sample.

The porosity of the samples was calculated using the relation

$$P=\left(\frac{D_x - D}{D_x}\right)x100\% \qquad (4)$$

The composition dependence of apparent density D, X-ray density D_x and porosity P is shown in the Table1 for the three series.

The increase in density with rare earth content can be attributed to the atomic weight and density of rare earths (138.9, 6.15 g/cm^3 ;144.2, 7.01g/cm^3 and 150.3, 7.52 g/cm^3) for La^{3+}, Nd^{3+} and Sm^{3+} respectively, which are higher than those of Sr (87.6, 2.54g/cm^3). The replacement of RE^{3+} by Sr^{2+} ions

Table 1 X-ray density D_X, Observed density D and Porosity P (%) of $Sr_{1-x}RE_xFe_{12}O_{19}$ where RE = La^{3+}, Nd^{3+} and Sm^{3+} with (x = 0 to 0.30)

RE^{3+}	Composition (x)	D (g/cm^3)	D_x (g/cm^3)	P (%)
La	0	4.20	5.07	17.03
	0.10	4.27	5.10	16.20
	0.20	4.31	5.13	16.02
	0.30	4.37	5.20	15.80
Nd	0.10	4.28	5.11	16.26
	0.20	4.32	5.15	16.03
	0.30	4.39	5.21	15.81
Sm	0.10	4.28	5.12	16.31
	0.20	4.33	5.16	16.07
	0.30	4.39	5.22	15.97

in the hexagonal structure leads to a variation in the bonding and consequently interatomic distance and density. The oxygen ions which diffuse through the material during sintering also accelerate the densification of the material. The apparent density of the same sample reflects the same general behavior of the theoretical density D_x. The X-ray density is higher than the apparent density value due to the existence of pores which depends on the sintering condition. It was determined by Archimedes principle based method .The porosity decreases as rare earth content increases reflecting the opposite behavior of density. The higher values of X-ray densities for Sm^{3+} ions substituted hexaferrites as compared to those for Nd^{3+} ions and La^{3+} ions substituted ones may be due to the lower value of lattice constants in former type of substitution.

Frequency variation of dielectric constant and dielectric loss tangent

The variation of the dielectric constant as a function of frequency at a constant temperature of 304 K for three series prepared with RE = La^{3+}, Nd^{3+} and Sm^{3+} are shown in figure 4. It is observed that the value of dielectric constant decreases with increasing frequency. This behaviour in rare earth substituted Sr hexaferrites is common ferrimagnetic behaviour and has also been observed by other investigators [16-18].

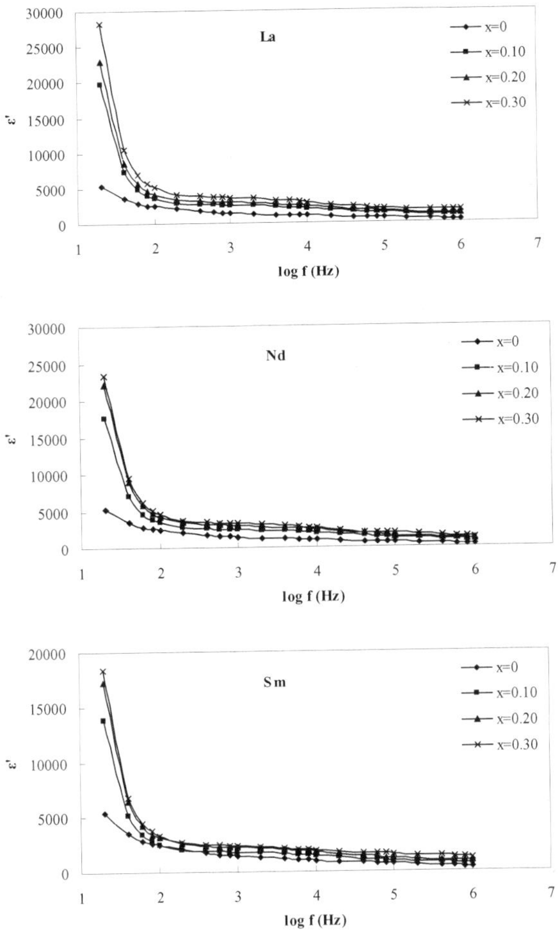

Figure 4. Variation of dielectric constant (ε′) with frequency at different compositions.

A more dielectric dispersion is observed at lower frequency range and it remains almost independent of applied external field at high frequency domain. The dielectric dispersion observed at lower frequency range is due to Maxwell -Wagner type interfacial polarization well in agreement with the Koops phenomenological theory [19, 20]. According to these models, the dielectric material with a heterogeneous structure can be imagined as a structure consists of well conducting grains separated by highly resistive thin layers (grain boundaries). In this case, the applied voltage on the sample drops mainly across the grain boundaries and space charge polarization is built up at the grain boundaries. The space charge polarization is governed by the available free charges on the grain boundary and the conductivity of the sample. Koops proposed that the effect of grain boundaries is predominant at low frequencies. The thinner the grain boundary, the higher the dielectric constant value is. The observed decrease of ε' with increasing the frequency can be attributed to the fact that the electron exchange between Fe^{2+} and Fe^{3+} ions can not follow the change of the external applied field beyond a certain frequency. Also it is observed that the dielectric constant increases with increase in x. Similar type of results have been reported in different ferrite [23, 24]. Both the dielectric constant and electrical conductivity are basically electrical properties and it has been recognized that the same mechanism viz. exchange of electrons between Fe^{2+} and Fe^{3+} are responsible for both the phenomena. A strong correlation between conduction mechanism and dielectric behaviour of ferrites has been established by Iwauchi [25] and Rezlescu and Rezlescu [26]. It has been concluded that the electron exchange between $Fe^{2+} \leftrightarrow Fe^{3+}$ results in the local displacement of charges, and this is responsible for polarization in ferrites. The magnitude of exchange, which also controls the conduction in ferrites, depend upon the concentration of Fe^{3+} / Fe^{2+} ion pairs present on B sites.

The Fe^{2+} ion concentration is a characteristic property of a given material and its value depends on several factors, sintering temperature, sintering atmosphere and annealing time, etc. Some amount of Fe^{2+} ions can also be formed due to partial reduction of Fe^{3+} ions during sintering. Therefore, the increase of dielectric constant with composition can be attributed to the excess formation of Fe^{2+} ions. Thus, it is the number of ferrous ions on the octahedral sites that play a dominant role in the process of conduction as well as dielectric polarization. At the same substitution content, the dielectric constant (ε') of La-substituted strontium hexaferrite is larger than those of Nd-doped and Sm-doped hexaferrites due to larger concentration of Fe^{3+} / Fe^{2+} ion pairs present on the octahedral sites. It has been observed that the dielectric constant (ε') increase with increase in x. The increase is, however, smaller in case of Nd and Sm substituted series for x= 0.30. The reason for this might be the existence of an insulating secondary phase on grain boundaries, which may reduce the Fe^{2+} ions on the octahedral sites and also it is true that extra phase at grain boundaries limits the reduction of Fe^{2+} ions concentration.

The variation of the dielectric loss tangent as a function of frequency at a constant temperature of 304 K for three series prepared with RE = La^{3+}, Nd^{3+} and Sm^{3+} are shown in figure 5. These figures explicitly show that the value of dielectric loss tangent increases as the frequency increases to attain its maximum (peak) value at a certain critical frequency followed by decrease at higher frequencies. The peaking behaviour in the dielectric loss occurs when the jump frequency of electron between Fe^{2+} and Fe^{3+} is equal to the frequency of the applied field [21, 22]. Also it is observed that the peaks are shifted towards higher frequencies accompanied with an increase in height of the peak with increasing RE ion substitution. The substitution of rare earth ions some how decreases the distance between Fe^{3+} and Fe^{2+}, resulting in a change in hopping frequency. The observed peaks of tan δ curves can be explained according to the fact that a strong correlation between the conduction mechanism and the dielectric behaviour exist in ferrites [27, 28].

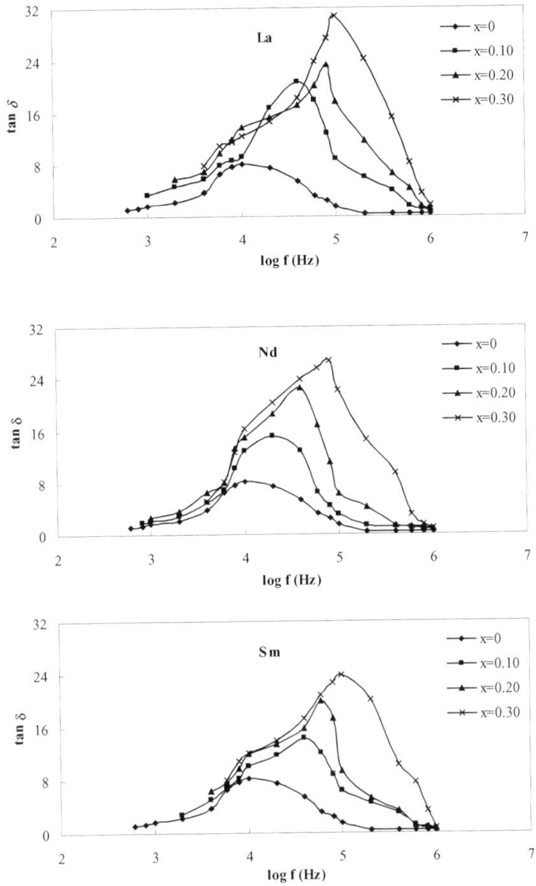

Figure 5. Variation of dielectric loss (tan δ) with frequency at different compositions.

In this case, the peak is expected when the hopping frequency of the electron between Fe^{2+} and Fe^{3+} ions is approximately equal to that of the external applied electrical field.
In this case

$$\omega\tau=1 \tag{5}$$

where τ is the relaxation time of the hopping process and ω is the angular frequency of the external field ($\omega = 2\pi f_{max}$). It is also known that the relaxation time τ is inversely proportional to the jumping probability per unit time, P, according to the relation [27]

$$\tau = \frac{1}{2} P \qquad (6)$$

So, from equations (5) and (6), it is expected that f_{max} is proportional to P. The shift and increase in height of the peak with increasing RE ion substitution indicates that the jumping probability per unit time, P increases as RE ion content increases. The increase of jumping probability may be attributed to the increase of ferrous ions in B-site, which is responsible for the polarization in these ferrites. The dielectric loss (tan δ) is observed to be maximum in case of La-substituted strontium hexaferrite. These losses in ferrites are generally reflected in the resistivity measurement. Materials with high resistivity (low conductivity) exhibit low dielectric losses and vice versa [29]. Increase of loss tangent values in RE-substituted Sr hexaferrites confirms the increase in conductivity [30, 31] supporting the Verway conduction mechanism [32].

Curie Temperature

Figure 6 shows the variation of the Curie temperature $T_c(K)$ with composition for all the series. It is observed that the value of Curie temperature $T_c(K)$ decreases with increasing RE content in each case. This behaviour is based on the number of magnetic ions present in the two sub-lattices and their mutual contraction due to the substitution of ions. It is known for SrM hexaferrite that the magnetic moments of Fe^{3+} ions arranged collinear due to the existence of superexchange interaction.

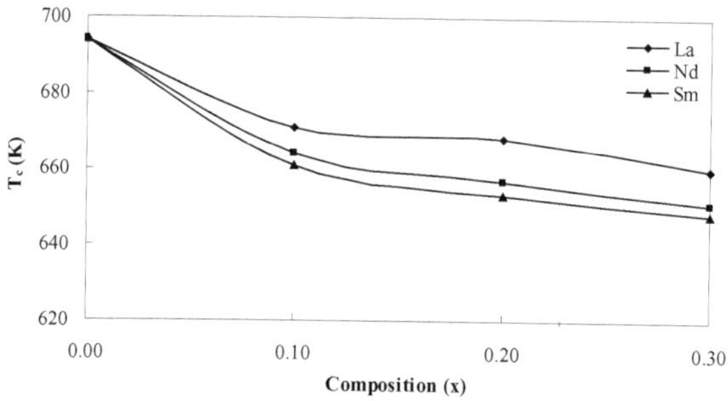

Figure 6. Variation of Curie temperature (T_c) with composition (x).

That is to say that it is the superexchange interaction to determine the orientation of magnetic moments of Fe^{3+} ions. Substitution of Sr^{2+} ions by RE^{3+} ions associated with a valence change of one Fe^{3+} ion per formula unit to Fe^{2+} ion can reduce the strength of this interaction due to Fe^{2+} ions. Since the Curie temperature $T_c(K)$ is determined by the overall strength of the exchange interactions, the weakening of exchange interactions result in decrease of Curie temperature as concentration of RE ions is increased. However for the same substitution content, the $T_c(K)$ value of La-doped strontium hexaferrite is correspondingly higher than those of Nd-doped ferrites and Sm-doped ferrites. It may be due to the small separation between the spin inside the domain and hence the exchange interaction will be large for La-doped sample as compared to the Nd and Sm doped samples.

CONCLUSIONS

M-type hexagonal ferrites $Sr_{1-x}RE_xFe_{12}O_{19}$ samples where RE = La^{3+}, Nd^{3+} and Sm^{3+} with (x = 0 to 0.30) were prepared by ceramic processing technique. The increase in dielectric constant and dielectric loss tangent with frequency has been explained with the Maxwell–Wagner type interfacial polarization in agreement with the Koops phenomenological theory. Out of the three series prepared, La-substituted strontium hexaferrite showed the larger value of dielectric constant (ε′) and dielectric tangent loss (tan δ) with increase in frequency. This variation may be attributed to the excess formation of Fe^{2+} ions which play a dominant role in the process of conduction as well as in dielectric polarization. Curie temperature decreases as rare earth ions substitution increases. This behaviour is based on the number of magnetic ions present in the two sub-lattices and their mutual contraction due to the substitution of rare earth ions.

REFERENCES

[1]G. K. Thompson and B. J. Evans, The structure–property relationships in M-type hexaferrites: Hyperfine interactions and bulk magnetic properties, *J. Appl. Phys.*, **73,** 6295-97, (1993).

[2]Z. W. Li, C. K. Ong, Z. Yong, F. L. Wie, X. Z .Zhou, J. H. Zhao and A. H. Morrish, Site preference and magnetic properties for a perpendicular recording material: $BaFe_{12-x}Zn_{x/2}Zr_{x/2}O_{19}$ nanoparticles, *Phys. Rev. B,* **62,** 6530- 37, (2000).

[3]A. Ataie and S. Heshmati-Manesh, Synthesis of ultra-fine particles of strontium hexaferrite by a modified co-precipitation method, *J. Eur. Ceram. Soc.*, **21,** 1951-55, (2001).

[4]O. Kubo, T. Nomura, T. Ido and H. Yokoyama, Improvement in the temperature coefficient of coercivity for barium ferrite particles, *IEEE Trans. Magn.*, **24,** 2859 – 61, (1988).

[5]Z. B. Guo, W. P. Ding, W. Zhong, J.R. Zhang and Y.W. Du, Preparation and magnetic properties of $SrFe_{12}0_{19}$ particles prepared by the salt-melt method, *J. Magn. Magn. Mater.*, **175,** 333-36, (1997).

[6]J. Barsik, Z. Simsa, L. Stichaver and R. Tesat, Magneto-optical properties of Co- and Ti-substituted hexagonal ferrite films prepared by the sol-gel method, ***J. Magn. Magn.****Mater.*, **157,** 311-12, (1996).

[7]F. Haberey and A. Kockel, The formation of strontium hexaferrite $SrFe_{12}O_{19}$ from pure iron oxide and strontium carbonate, *IEEE Trans. Magn.*, **12,** 983- 985, (1976).

[8]G. Albanese, B. E. Watts, F. Leccabue and S. D. Castanon, Mossbauer and magnetic studies of $PbFe_{12-x}Cr_xO_{19}$ hexagonal ferrites, *J. Magn. Magn. Mater.* **184,** 337-43, (1998).

[9]K. Kakizaki, N. Hirasuta and Y. Namikawa, Fine structure of acicular $BaCo_xTi_xFe_{12-2x}O_{19}$ particles and their magnetic properties, *J. Magn. Magn. Mater.*, **176,** 36-40, (1997).

[10]L. Lechevallier, J. M. Le Breton, A. Morel and J. Teillet, Structural and magnetic properties of $Sr_{1-x}Sm_xFe_{12}O_{19}$ hexagonal ferrites synthesised by a ceramic process, *J. Alloys and Compd.*, **359,** 310-14, (2003).

[11]H. Mocuta, L. Lechevallier, J. M. Le Breton, J. F. Wang and I. R. Harris, Structural and magnetic properties of hydrothermally synthesised Sr-Nd hexagonal ferrite, *J. Alloys and Compd.*, **364,** 48-52, (2004).

[12]A. Singh, S.B. Narang, K. Singh, P. Sharma and O.P. Pandey, Structural, AC conductivity and dielectric properties of Sr-La hexaferrite, *Eur. Phys. J. Appl. Phys.*, **33,** 189-193, (2006).

[13]R.F. Sahoo, *Theory and applications of ferrites,* Prentice Hall, p109-110, 1960.

[14]X. Liu, W. Zhong, S.Yang, Z. Yu, B. Gu , Y. Du, Influences of La^{3+} substitution on the structure and magnetic properties of M-type strontium ferrites, *J. Magn. Magn. Mater.*, **238,** 207-14, (2002).

[15]S. Ounnunkad, Improving magnetic properties of barium hexaferrites by La or Pr substitution, *Solid Stat. Commn.*, **138,** 472-475, (2006).

[16]Ravinder, D., Vijaya, P and Reddy, B., High-frequency dielectric behaviour of Li–Mg ferrites, *Mater. Lett.*, **57,** 4344 -50, (2003).

[17]Shaikh, A.M., Belled, S.S and Chougule, B.K., Temperature and frequency-dependent dielectric properties of Zn substituted Li–Mg ferrites, *J. Magn. Magn. Mater.*, **195,** 384 -90,

(1999).
[18]B.R. Kumar, D. Ravinder, Dielectric properties of Mn–Zn–Gd ferrites, *Mater. Lett.*, **53,** 437-40, (2002).
[19]C.G. Koops, On the Dispersion of Resistivity and Dielectric Constant of Some Semiconductors at Audiofrequencies, *Phys. Rev.*, **83,** 121-24, (1951).
[20]K.W. Wagner, *Ann. Phys.*, **40,** 817-19, (1913).
[21]M.B. Reddy, P.V. Reddy, Low-frequency dielectric behaviour of mixed Li-Ti ferrites, *J. Appl. Phys. D,* **24,** 975-81, (1991).
[22]A. M. Abo El Ata, S.M. Attia, and T.M. Meaz, AC conductivity and dielectric behavior of $CoAl_xFe_{2-x}O_4$, *Solid Stat. Sci.*, **6,** 61- 69, (2004).
[23]N. Kumar, P. Kishan and Z.H. Zaidi. Effect of MgTi and ZnTi substitutions on electrical and magnetic properties of Li ferrite, *J. Magn. Magn. Mater.* **184,** 355- 57, (1998).
[24]S.S. Bellad and B.K. Chougle, Composition and frequency dependent dielectric properties of Li– Mg–Ti ferrites, *Mater. Chem. and Phys.*, **66,** 58-63, (2000).
[25]K. Iwauchi, Dielectric Properties of Fine Particles of Fe_3O_4 and Some Ferrites, Jpn. *J. Appl. Phys.*, **10,** 1520-28, (1971).
[26]N. Rezlesu, E. Rezlescu, Dielectric properties of copper containing ferrites, Phys. Stat. Sol. (A), **23,** 575-82, (1974).
[27]M.B. Reddy and P.V. Reddy, Low-frequency dielectric behaviour of mixed Li-Ti ferrites, *J. Appl. Phys. D,* **24,** 975-81, (1991).
[28]S.C. Watawe, B.D. Sarwede, S.S. Bellad, B.D. Sutar and B.K. Chougule, Microstructure, frequency and temperature-dependent dielectric properties of cobalt-substituted lithium ferrites, *J. Magn. Magn. Mater.*, **214,** 55-60, (2000).
[29]C.B. Kolekar, P.N. Vasambekar, S.G. Kulkarni and A.S. Vaingankar, Effect of Gd^{3+} substitution on dielectric behaviour of copper-cadmium ferrites, *J. Mater. Sci.*, **30,** 5784-88, (1995).
[30]N.J. Ali, J. Rahman and M.A. Showdhaury, The Influence of the Addition of CaO on the Magnetic and Electrical Properties of Ni–Zn Ferrites, *Jpn. J. Appl. Phys.*, **39,** 3378- 81, (2000).
[31]A.Y. Lipare, P. N. Vasambekar and A. S. Vaingankar, X-ray, IR and dc electrical resistivity study of $CaCl_2$-doped zinc-copper ferrite system, *Phys. Stat. Sol. (A),* **196,** 372-78, (2003).
[32]E.J.W. Verway, P.M. Haaijman, G.M. Romeyn and F.C. Vas Oosterhout, Physical properties and cation arrangements of oxides with spinal structure, *J. Chem. Phys.*,**15,** 181-87, (1947).

HIGH TEMPERATURE PIEZOELECTRIC PROPERTIES OF SOME BISMUTH LAYER-STRUCTURED FERROELECTRIC CERAMICS

Tadashi Takenaka, Hajime Nagata, Toji Tokutsu, Kazuhiro Miyabayashi, and Yuji Hiruma
Faculty of Science and Technology, Tokyo University of Science,
Yamazaki 2641, Noda, Chiba, 278-8510, Japan

ABSTRACT

We focused on some BLSF ceramics with high Curie temperature, T_c such as $Bi_4Ti_3O_{12}$ (BIT), $Na_{0.5}Bi_{4.5}Ti_4O_{15}$ (NBT), $CaBi_4Ti_4O_{15}$ (CBT4) and $CaBi_2Ta_2O_9$ (CBT2), and examined their piezoelectric properties at room temperature. As results, $Bi_4Ti_{0.98}V_{0.02}O_{12}$ (BITV-0.02) indicated relatively high Tc of 678°C and the largest piezoelectric voltage constant, g_{33}, of 22×10^{-3} Vm/N among above compounds. The grain oriented BITV-0.02 indicated large g_{33} ~35×10^{-3} Vm/N at room temperature and g_{33} ~27×10^{-3} Vm/N at 400°C from the in-situ measurement of resonance and antiresonance method. Therefore, the BITV-0.02 ceramic seems to be good candidate for lead-free high-temperature piezoelectric materials.

INTRODUCTION

The industrial and scientific communities have expressed the need for sensing over a broad temperature range. The maximum operation temperature of piezoelectric materials for sensors is limited by Curie temperature, T_c, combined with the level of sensitivity determined by the piezoelectric voltage constant, g_{ij}. Bismuth layer-structured ferroelectrics (BLSFs)[1-3] are attractive from the viewpoint of sensor applications, because BLSFs are characterized by their low dielectric constant, ε_s, high Curie temperature, T_c, and large anisotropy in the electromechanical coupling factor k_t/k_p or k_{33}/k_{31}[4-6]. For example, $CaBi_2Ta_2O_9$ (CBT2), is expected to be a candidate material for high temperature sensors due to its high-T_C, approximately 900°C[7]. However, there are few reports about piezoelectric properties for CBT2-based ceramics due to the difficulty in poling. In this study at first, we focused various BLSF compounds, such as $CaBi_2Ta_2O_9$ (CBT2), $Bi_4Ti_3O_{12}$ (BIT), $Na_{0.5}Bi_{4.5}Ti_4O_{15}$ (NBT), and $CaBi_4Ti_4O_{15}$ (CBT4) based ceramics as the high Tc material and we tried to clarify their piezoelectric properties at room temperature and to compare these properties.

In terms of $Bi_4Ti_3O_{12}$ (BIT), it is a typical well-known BLSF with intermediate T_c of 680°C[8-11]. However, it is difficult to measure the piezoelectric properties of BIT ceramics due to a difficulty in poling because of low resistivity. To solve these problems, V^{5+} ions were doped into a BIT ceramic to obtain higher resistivity and to enhance the piezoelectricity[9-12]. We have already reported that V doping into the BIT ceramic was very effective to enhance the piezoelectric properties, and $Bi_4Ti_{0.98}V_{0.02}O_{12}$ [BITV-0.02] ceramic showed the largest electromechanical coupling factor k_{33} of 0.25 and high resistivity >10^{12} Ω·cm at RT[13]. Therefore, BITV ceramic seems to be a good candidate material for high temperature sensors with intermediate T_c. In the second part of this study, the grain orientation effects of $Bi_4Ti_{0.98}V_{0.02}O_{12}$ [HF-BITV-0.02] ceramics were investigated using the hot-forging (HF) method for enhancing the piezoelectric properties and the high temperature piezoelectric characteristics were examined for the high temperature sensor applications.

EXPERIMENTAL PROCEDURE

Starting raw materials for CBT2, BITV, NBT and CBT4 ceramics were Bi_2O_3, Ta_2O_5, TiO_2, $CaCO_3$, V_2O_5 and Na_2CO_3 of purity higher than 99.9%. These materials were mixed by ball milling and calcined at 800-850°C for 2-4 h. After the calcination, the ground and ball-milled powders were pressed at 150 MPa and sintered at 950-1310°C for 2-4 h.

Grain-oriented samples for BITV-0.02 were prepared by the hot-forging method (HF)[8-9]. Starting raw materials were mixed by ball milling and calcined at 600°C for 1 h and 850°C for 2 h. After the calcination, the ground and ball-milled powders were pressed into disks 20 mm in diameter and about 30 mm in thickness. Figure 1 (a) shows a drawing of the HF apparatus. The pre-pressed cylinder (green body) was sandwiched between two platinum plates to prevent the sample from reacting with alumina plate. A uniaxial compression of 100-200 kg/cm^2 was statically applied along the thickness of the sample (the forging axis). A hot-forging program is shown in Fig. 1 (b). After conventional firing for 1 hour, the sample was gradually pressed by two alumina plungers for 2 hours. Then a constant pressure and temperature were maintained for 2 hours. The maximum temperatures were at 920-960°C. The grain orientation factor, F, was calculated using the Lotgering method[14]. The phases of the HF samples were identified by X-ray diffraction (XRD) analysis. Samples were polished and thermally etched for microstructural examination by scanning electron microscopy (SEM, HITACH S-2400).

The sintered ceramics were cut and polished into rectangular specimen of 4×2×2 mm^3 to determine the piezoelectric properties of the (33) mode. Specimens for piezoelectric measurements were poled in stirred silicone oil at an applied field of E_p=5-7 kV/mm, a temperature of T_p=200-280°C, and a time of t_p= 5-30 min. Piezoelectric properties were measured by a resonance-antiresonance method on the basis of IEEE standards, using an

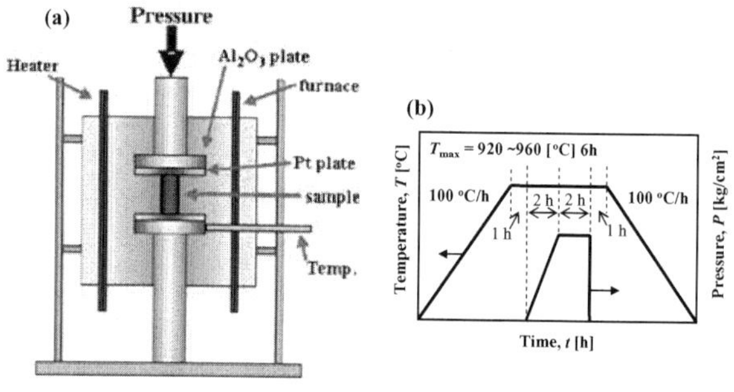

Figure 1. Hot-forging apparatus, (a), and program, (b).

impedance analyzer (YHP 4192A and HP 4294A). The electromechanical coupling factor, k_{33}, was calculated from the resonance and antiresonance frequencies. The free permittivity, ε_{33}^{T}, was determined from the capacitance at 1 kHz of the poled specimen. The elastic constants, s_{33}^{E}, were calculated from the frequency constant, N_{33}, and the measured density, ρ_0. Finally, the piezoelectric strain constants, d_{33}, and piezoelectric voltage constant, g_{33}, were calculated from k_{33}, ε_{33}^{T} and s_{33}^{E}. The temperature dependences of k_{33} were measured in the temperature range from RT to 600°C using the resonance-antiresonance method.

RESULTS AND DISCUSSION

Piezoelectric properties of some BLSF compounds at room temperature

In this section, we mainly explain the piezoelectric properties at room temperature of CBT2, BITV, NBT and CBT4 ceramics. XRD patterns of all ceramics showed a single phase of bismuth layer-structured compounds and the relative density to the theoretical density was higher than 90%. Resistivities, ρ, for CBT2, BITV, NBT and CBT4 ceramics were higher than 10^{12} Ω·cm at room temperature. Therefore, all ceramics prepared in this study were expected to be poled under a high electric field during a poling treatment. Figure 2 shows the temperature dependences of dielectric constant, ε_S, and at 1-30 MHz for the CBT2, BITV, NBT and CBT4 ceramics. It indicated that the Curie temperatures, T_c, of these ceramics showed higher than 600°C, especially the CBT2 indicated the highest T_c of 923°C.

Figure 3 shows applied field dependences of the coercive field, E_C, and the remanent polarization, P_r, for CBT2 ceramic at each temperature. The E_C and P_r start to increase at lower E_a with increasing temperature. Especially, the E_C at 250°C is saturated at about 170 kV/cm, which is the lowest E_a. Figure 4 shows temperature dependences of (a) E_C and P_r and (b) P-E

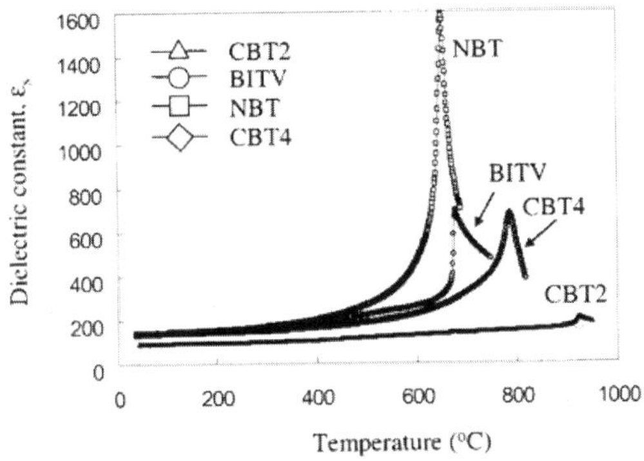

Figure 2. Temperature dependences of dielectric constant, ε_S, for $CaBi_2Ta_2O_9$ (CBT2), $Bi_4Ti_{2.98}V_{0.02}O_{12}$ (BITV), $Na_{0.5}Bi_{4.5}Ti_4O_{15}$ (NBT), and $CaBi_4Ti_4O_{15}$ (CBT4) ceramics

hysteresis loops at 50°C and 250°C for CBT2 ceramic. The P_r increases with increasing temperature and becomes saturated. On the other hand, E_C decreases with increasing temperature between 100°C and 250°C. However, between 50°C and 100°C, the E_C increases with increasing temperature. The reason is assumed that the domain wall motions are limited at lower temperature. The *P-E* hysteresis loop is not so sharp at 50°C in Fig. 4(b) because of difficulties in inversion of domains. Therefore, it is expected that higher temperatures than 250°C could promote the domain wall motion during the poling treatment.

To obtain the well poled samples for CBT2 ceramic, samples were poled under various conditions. Figure 5 shows applied field dependences of (a) electromechanical coupling factor,

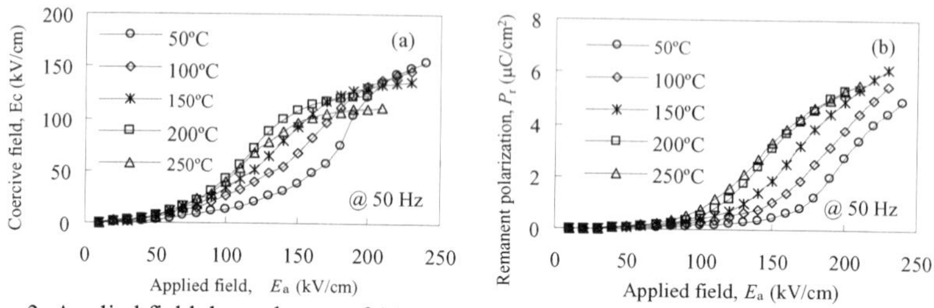

Figure 3. Applied field dependences of (a) coercive field, E_c, and (b) remanent polarization, P_r, for CBT2 ceramic at each temperature.

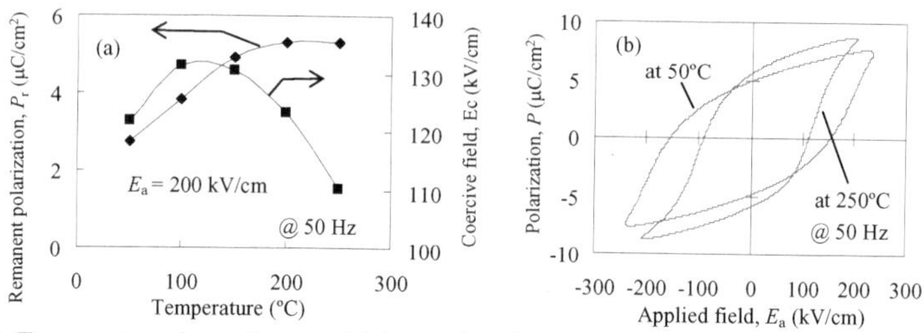

Figure 4. Temperature dependences of (a) coercive field, E_c, and remanent polarization, P_r, and (b) *P-E* hysteresis loops for CBT2 ceramic at 50°C and 250°C.

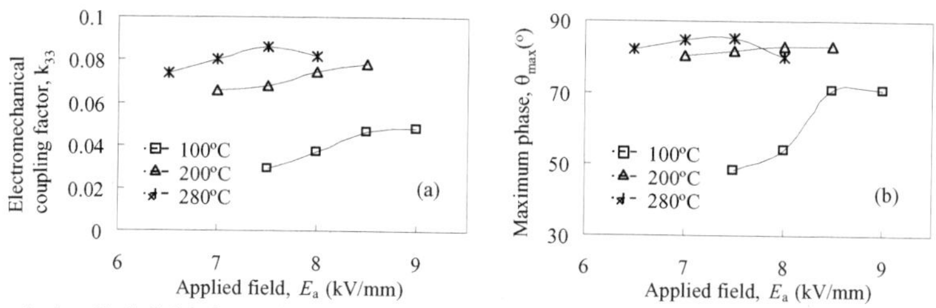

Figure 5. Applied field dependences of (a) electromechanical coupling factor, k_{33} and (b) the maximum phase, θ_{max}, for CBT2 ceramic in silicone oil heated at 100, 200 and 280°C.

k_{33}, and (b) maximum phase, θ_{max} at 100°C, 200° and 280°C for 10 min. The maximum temperature was restricted at 280°C due to the limitation of the silicone oil (Shin-Etsu Chemical Co. Ltd. KF-54). DC fields were applied to the samples until the breakdown. Both (a) and (b) indicate that the θ_{max} and k_{33} increase with increasing poling temperature and applied field. Especially, these values were drastically improved by increasing temperature as we expected. The largest values of the θ_{max} and the k_{33} at 280°C are 85.3° and 0.086 (at 7.5 kV/mm), respectively. The sample poled at 8 kV/mm at 280°C showed slightly lower values than those at 7.5 kV/mm. The θ_{max} and the k_{33} were deteriorated because the sample poled at 8 kV/mm and 280°C showed partially breakdown. From these results, the best applied field and temperature for poling are determined to 7.5 kV/mm and 280°C. Then, to determine the best poling time, samples were poled with increasing poling time. Figure 6 shows poling time dependences of the θ_{max} and k_{33} under the conditions of E_p = 7.5 kV/mm and T_p = 280°C. This result indicates that alignments of spontaneous polarization in ceramic grains are progressing with increasing poling time until 10 min. However, in the case of poled for longer than 20 min, the θ_{max} and the k_{33} are almost constant. Therefore, the best poling conditions are E_p = 7.5 kV/mm, T_p = 280°C and t_p = 20 min, respectively, for this sample. Figure 7 shows frequency dependences of the impedance, $|Z|$, and the phase, θ, on the (33) mode for CBT2 ceramic poled at the best condition. The sample poled under this condition showed good resonance characteristics with θ_{max} = 86.1°.

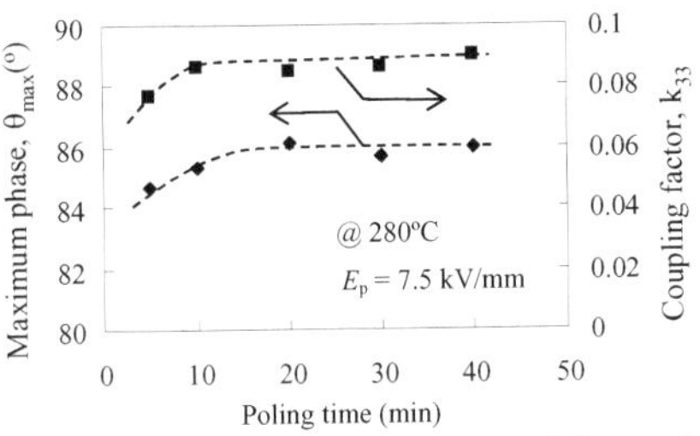

Figure 6. Poling time dependences of θ_{max} and the k_{33} for CBT2 ceramic.

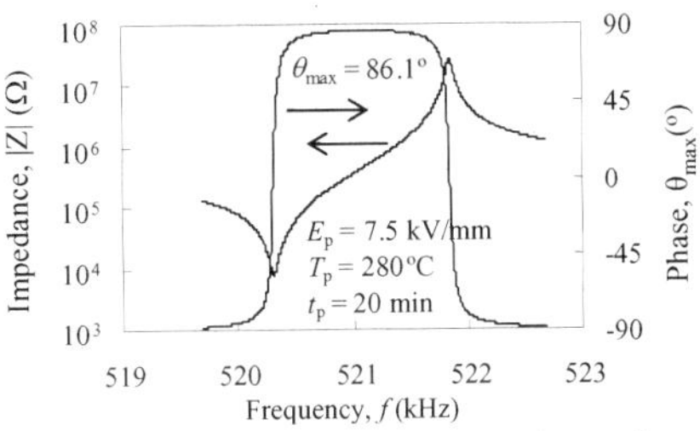

Figure 7. Frequency dependences of impedance, $|Z|$, and the phase, θ, on the (33) mode for CBT2 ceramic.

Table I Electrical and piezoelectric properties at room temperature for CBT2, BITV-0.02, NBT and CBT4 ceramics

Material	T_c (°C)	ε_r	d_{33} (pC/N)	g_{33} (10^{-3} Vm/N)	k_{33}	r (10^{12} Ωcm)
CBT2(m=2)	923	81	5.4	8.2	0.09	100
BITV-0.02	679	138	26	22	0.27	1
NBT	654	140	18	15	0.11	1000
CBT4(m=4)	788	139	13.9	12.5	0.15	1000

Table I summarizes electrical and piezoelectric properties at room temperature for CBT2, BITV-0.02, NBT and CBT4 ceramics. From Table I, piezoelectric properties for CBT2 ceramic are not so higher than those of other BLSF ceramics. However, this work confirmed that the CBT2 ceramic has the high T_C material and clarified its piezoelectric properties for the first time. On the other hand, BITV-0.02 ceramic indicates the largest g_{33} value among the BLSF ceramics in this study.

Textured $Bi_4Ti_{0.98}V_{0.02}O_{12}$ (BITV-0.02) ceramic

The sintering temperature of 960°C is low enough to suppress the Bi vaporization. The TG analysis was performed to confirm the suppression of the Bi vaporization, showing no weight loss up to 1000°C. X-ray diffraction patterns of HF-BITV-0.02 ceramics showed a single phase of bismuth layer-structured compounds (m=3) with orthorhombic symmetry. The relative densities of HF-BITV ceramics are higher than 95%. Orientation factors, F, determined from the X-ray diffraction patterns of the HF-BITV-0.02 were approximately 80-90%. The HF ceramic

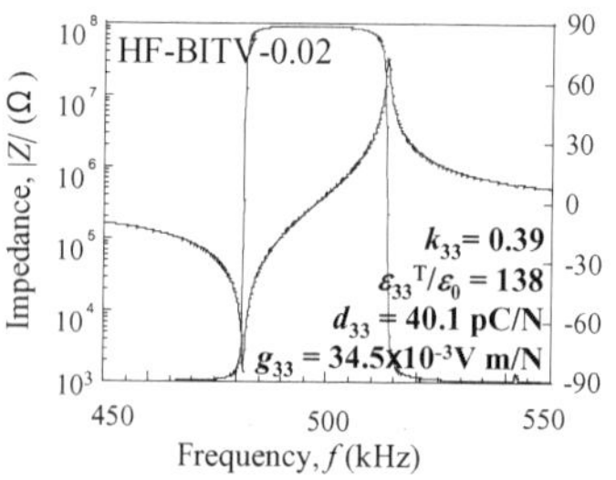

Figure 8. Frequency characteristics of impedance, Z, of the HF-BITV-0.02 ceramic at room temperature.

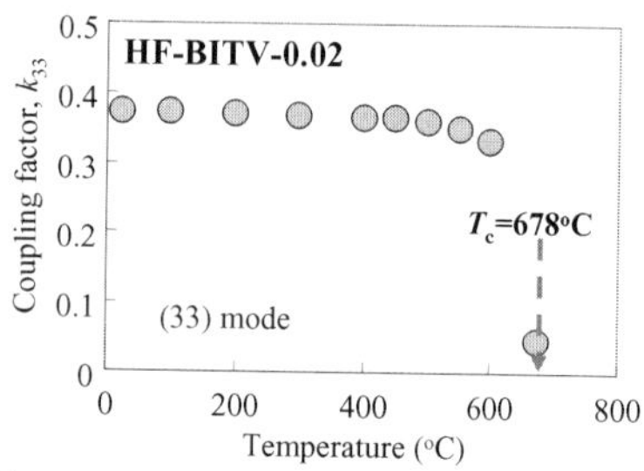

Figure 9. Electromechanical coupling factors, k_{33} as a function annealing temperature.

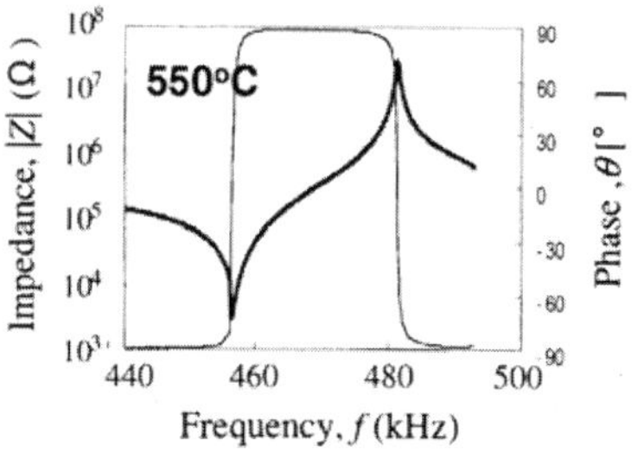

Figure 10. Frequency characteristics of impedance, Z, of the HF-BITV-0.02 ceramic at room temperature after the annealing at 550°C for 10 min.

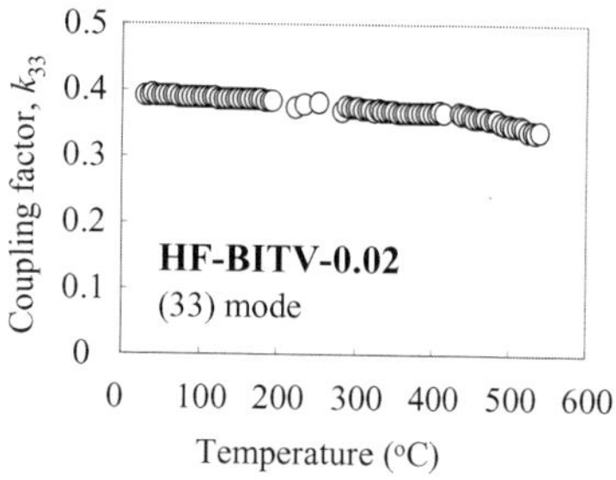

Figure 11. In-situ temperature dependence of electromechanical coupling factor, k_{33} for HF-BITV-0.02 ceramic.

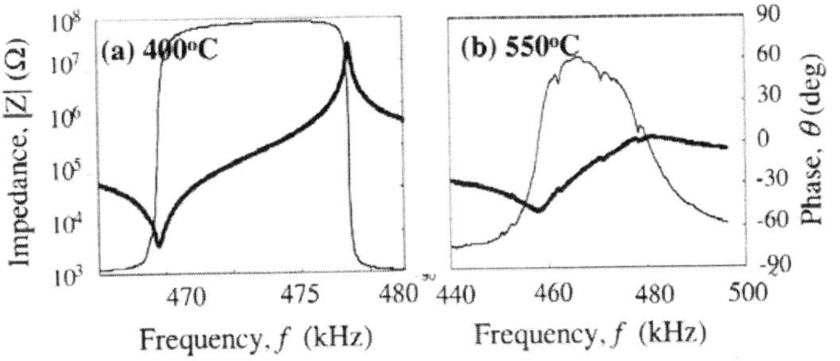

Figure 12. Frequency characteristics of impedance, Z, of the HF-BITV-0.02 ceramic at (a) 400°C and (b) 550°C.

exhibits a unique microstructure in which plate-like grains were aligned along the perpendicular direction for the forging axis from the SEM observation.

Figure 8 shows the frequency dependences of the impedance, |Z|, and phase, θ, for HF-BITV-0.02. The optimum poling conditions for the ceramic were 7 kV/mm, 200°C and 7 min. When we poled this ceramic at 250°C, it was difficult to apply high electric field more than 6 kV/mm due to lowering the resistivity. The maximum phase angle, θ_{max}, in the inductance (L) region between resonance and antiresonance frequency reached 88.7°, which is very close to 90°. The θ_{max} value is one of the standard parameters for indicating the degree of full poling. Thus, the θ_{max} of 88.7° means that the HF-BITV ceramic was poled sufficiently. Piezoelectric properties of k_{33}, d_{33}, g_{33} and Q_m were described in this figure, indicating 0.39, 40.1 pC/N, 34.5x10^{-3} Vm/N and 1500, respectively. The k_{33} value is about twice as large as that of non-oriented one, and is larger than that of previous report[3] for HF BIT ceramics (k_{33}=0.27). When we compared with other perovskite-type materials[15-20], PZT, $PbTiO_3$ (PT) and PT-$BiScO_3$ indicate high g_{33} values of approximately 30x10^{-3} Vm/N. However, T_cs of these materials are relatively low about 350-470°C. On contrary, BLSFs and $LiNbO_3$ (LN) exhibit high T_cs but g_{33} values of those materials are relatively small. The HF-BITV-0.02 in this study has both high T_c and high g_{33} values, which is very good sign for the high temperature sensor applications.

The thermal-depoling behavior of HF-BITV-0.02 ceramic is shown in Figure 9, in which k_{33} values are plotted against the annealing temperature. This experiment were conducted by holding at each high temperature for 10 min, cooling room temperature, measuring resonance and antiresonance frequencies and repeating the procedure up to 675°C. The k_{33} value was stable up to 600°C and dropped rapidly above 650°C near the T_c of 678°C. Figure 10 shows the frequency dependence of the impedance, |Z|, and phase, θ, on HF-BITV-0.02 annealed at 550°C. Good resonance and antiresonance wave is still maintained after the annealing at 550°C. The k_{33} and Q_m keep high values of approximately 0.37 and 1300, respectively, without a depolarized degradation.

Figure 11 shows the *in-situ* temperature dependence of k_{33} of HF-BITV-0.02 for the temperature range from RT to 550°C. At a glance of this figure, the k_{33} maintained up to 550°C as same as the thermal-depoling behavior in Fig. 9. However, if we look at the resonance and antiresonance wave in Figure 12(b), the θ_{max} is quite decrease to 60° and the impedances, |Z|, at resonance and antiresonance frequencies are obviously degraded. Figure 13 shows the temperature dependence of θ_{max} and Q_m of HF-BITV-0.02. Both θ_{max} and Q_m decrease rapidly above 450°C. From the result of the thermal-depoling behavior, this degradation is not because of the depolarization. It may be due to a rise of dielectric loss, tan δ, and conductivity. In the sensing devices, no electric field applies to the piezoelectric element, but a large RC time constant is required to detect the signals. Thus, further development and studies about the resistivity for this material are necessary. Anyhow, Figure 12(a) at high temperature of 400°C shows a good resonance and antiresonance wave, and the k_{33} value is 0.37. Therefore, the HF-BITV-0.02 ceramics seem to be good candidate for lead-free high-temperature piezoelectric materials.

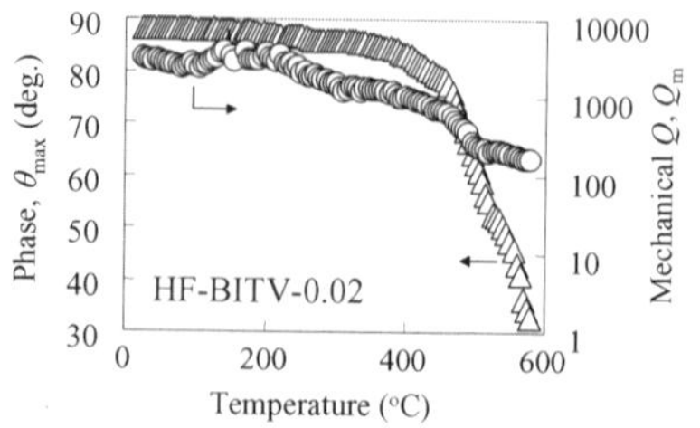

Figure 13. Temperature dependence of mechanical quality factor, Q_m, and maximum phase, θ_{max}, for HF-BITV-0.02

CONCLUSIONS

The temperature dependences of piezoelectric properties for grain oriented $Bi_4Ti_{2.98}V_{0.02}O_{12}$ [HF-BITV-0.02] ceramics are studied. Piezoelectric properties of electromechanical coupling factor, k_{33}, piezoelectric strain constant, d_{33}, piezoelectric voltage constant, g_{33}, and mechanical quality factor, Q_m, at room temperature were 0.39, 40.1 pC/N, 34.5x10^{-3} Vm/N and 1500, respectively. Considering the high Curie temperature, T_c, (=678°C), the HF-BITV-0.02 ceramics seem to be good candidate for lead-free high-temperature piezoelectric materials. We confirmed the thermal depoling behavior and piezoelectricity was stable up to 600°C. From the real time temperature dependences of piezoelectric properties, good resonance and antiresonance curve and the k_{33} higher than 0.37 maintained for the temperature range from RT – 400°C.

ACKNOWLEDGMENT

The authors would like to thank TOHO TITANIUM CO., LTD. for providing high purity titanium oxide powder. This work was partially supported by a Grant-in-Aid for Scientific Research (B) (No. 19360302) from the Japan Society for the Promotion of Science.

REFERENCES

[1]G. A. Smolenskii, V. A. Isupov and A. I. Agranovskaya, *Sov. Phys.-Solid State*, **3** [3] (1961), p. 651.
[2]E. C. Subbarao, *J. Am. Ceram. Soc.*, **45**, [4] (1962), p. 166.
[3]L. E. Cross and R. C Pohanka, *Mat. Res. Bull.*, **6** (1971), p.939.
[4]S. Ikegami and I. Ueda, *Jpn. J. Appl. Phys.*, **13** [10] (1974), p. 1572.
[5]T. Takenaka and K. Sakata, *Jpn. J. IEEE (C)*, **J65-C** (1982), p. 512. (in Japanese)
[6]H. Nagata, T. Takahashi and T. Takenaka, *Transactions of the Materials Research Society of Japan*, **25** [1] (2000) p. 273.

[7]K. Shibata, K. Shoji and K. Sakata, *Jpn. J. Appl. Phys.*, **40** [9B] (2001), p. 5719.
[8]S. E. Cummins and L. E. Cross, *J. Appl. Phys.*, **39** [5] (1968), p. 2268.
[9]T. Takenaka and K. Sakata, *Jpn. J. Appl. Phys.*, **19** [1] (1980), p. 31.
[10]M. Villegas, A. C. Caballero, C. Moure, P. Duran and J. F. Fernandez, *J. Am. Ceram. Soc.*, **82** [9] (1999), p. 2411.
[11]H. S. Shulman, M. Testorf, D. Damjanovic and Nava Setter, *J. Am. Ceram. Soc.*, **79** [12] (1996), p. 3124.
[12]Y. Noguchi and M. Miyayama, *Appl. Phys. Lett.*,Vol. **78** [13] (2001), p. 1903.
[13]H. Nagata, Y. Fujita, H. Enosawa and T. Takenaka, *Ceramic Transactions,* **150** (2004), p. 253.
[14]F. K. Lotgering, *J. Inorg. Nucl. Chem.*, **9** (1959), p. 113.
[15]J. M. Hervert, Ferroelectric trasducers and Sensors, Gordon and Breach Science Publishers, New York, 1982.
[16]Tokin Corp., Piezoelectric Ceramic Catalog, Tokin Corp, Tokyo, Japan.
[17]M. E. I. Corp., Product Catalog, Matsushita Electric Industrial Co., Osaka, Japan.
[18]I. Crystal Technology, Product Catalog, Crystal Technology Inc., Palo Alto.
[19]R. E. Eitel, C. A. Randall, T. R. Shrout, P. W. Rehrig, W. Hackenberger, and S. –E. Park, Jpn. J. Appl. Phys., **40**[10] (2001) p. 5999.
[20]H. Yan, H. Zhang, R. Ubic, M. J. Reece, J. Liu, Z. Shen and Z. Zhang, Advanced Materials, **17** (2005) p. 1261.

EFFECTIVE SIZE OF VACANCIES IN THE $Sr_{1-3x/2}Ce_xTiO_3$ SUPERSTRUCTURE

Rick Ubic
Boise State University
Boise, Idaho, USA

Ganesanpotti Subodh and Mailadil T. Sebastian
National Institute for Interdisciplinary Science and Technology
Trivandrum, India

Delphine Gout and Thomas Proffen
Los Alamos Neutron Science Cente
Los Alamos, New Mexico, USA

ABSTRACT

Four compositions in the $Sr_{2-3x/2}Ce_xTiO_3$ homologous series corresponding to $x = 0.1333$, 0.1667, 0.1818, 0.25, and 0.4 have been synthesized by mixed-oxide processing and crystallographically characterized by electron diffraction. A non-cubic perovskite superlattice is observed for each composition caused by antiphase octahedral tilting. The degree of tilt increases with increasing x. Structural refinements of neutron diffraction data previously published indicate that the A-site size decreases with increasing x, but not as much as would be expected if vacancies were truly zero-dimensional defects; instead, A-site vacancies posses an effective finite size, presumably due to mutual repulsion of the coordinating oxygen ions. Based on these results, modifications to the predictive formula for pseudocubic lattice constant are proposed and seem successful for several systems with disordered A-site vacancies.

INTRODUCTION

Perovskite oxides have the general formula ABO_3, and complex perovskites of the type $A(B'B'')O_3$ are now well-established materials for microwave applications because of their high quality factors (Qf) and low temperature coefficients of resonant frequency (τ_f). These materials typically contain divalent species on the A-site, and compositions with trivalent rare-earth cations doped onto the A-site are typically charge-compensated by the simultaneous incorporation of a trivalent species like Al^{3+} on the B-site, *e.g.*, $(Ca_{1-x}Nd_x)(Ti_{1-x}Al_x)O_3$. In the Ce-doped $SrTiO_3$ system, it has been shown[1,2] that charge compensation occurs simply via A-site vacancy formation, with Ce^{3+} dopants occupying Sr^{2+} sites and creating A-site vacancies via the reaction:

$$Ce_2O_3 \xrightarrow{SrTiO_3} 2Ce_{Sr}^{\bullet\bullet} + V_{Sr}'' + 3O_O^x \qquad (1)$$

The structure of $Sr_{1-3x/2}Ce_xTiO_3$ ($0.1333 \leq x \leq 0.4$) was recently reported as cubic, space group $Pm\bar{3}m$, on the basis of x-ray diffraction data[3] and vibrational spectroscopy work.[4] More recently, Ubic *et al.*[2] have shown by electron and neutron diffraction that the correct symmetry is lower than cubic, probably $R\bar{3}c$. Oxygen octahedra are tilted about the pseudocubic [111] by up to 4.7°.

The perovskite tolerance factor, the relationship of which to the pseudocubic perovskite lattice constant is discussed at some length in reference 5, is defined as:

$$t = \frac{r_A + r_O}{\sqrt{2}(r_B + r_O)} \tag{2}$$

where r_A, r_B, and r_O are the ionic radii of the ionic species on the A, B, and O sites, respectively. Reaney *et al.*[6] established a relationship between tolerance factor and the temperature coefficient of permittivity (τ_ε) in which it was determined that changes in τ_ε were closely correlated to octahedral tilt transitions in niobate and tantalate perovskites. They observed that perovskites with t < ~0.985 contained axes about which oxygen octahedra were tilted in an anti-phase arrangement, causing cell doubling in the three pseudocubic directions. Similarly, perovskites for which t < ~0.965 undergo a further tilt transition whereby octahedra are tilted in-phase about one or more axes as well. Perovskites for which t > ~0.985 were not observed to contain a tilt superlattice.

PROCEDURE

Stoichiometric amounts of $SrCO_3$ and TiO_2 (99.9%, Aldrich Chemical Co., Milwaukee, WI) and CeO_2 (99.99%, Indian Rare Earth Ltd., Udyogamandal, India) were ball milled in distilled water for 24 hours, using YSZ media in a plastic pot. Slurries were dried, ground, and calcined at 1100°C for five hours. Around 4 wt% of polyvinyl alcohol (PVA) (molecular weight 22000, 88% hydrolyzed, BDH Lab Supplies, England) was added to the dried powders, which were then ground again into fine powder. Cylindrical pellets of about 1-2 mm height and about 14 mm diameter were made by applying a pressure of 100 MPa. These compacts were then fired at 600°C for 30 min to burn the binder out before sintering for two hours at temperatures ranging from 1300 to 1400°C. Samples for transmission electron microscopy (TEM) were prepared by thinning pellets to electron transparency by conventional ceramographic techniques followed by ion thinning (XLA/2000, VCR Group, San Clemente, California, USA) to electron transparency for observation in the TEM (2100 HR, JEOL, Japan).

RESULTS

All the compositions processed were single-phase and dense, as already reported.[1,2] Fig. 1 shows <110> selected area electron diffraction patterns for several compositions in the series. With the exception of $x = 0$ ($SrTiO_3$), which is included for comparison, all the patterns show the presence of ½{*odd,odd,odd*} superlattice reflections (α-type) indicative of antiphase octahedral rotations about pseudocubic <100>. In the case of $x = 0.40$ (Fig. 1f), additional superlattice reflections are present with ½{*odd,even,even*} indices, possibly indicative of antiparallel A-site cation displacements; however, their slightly blurred and indistinct nature suggests that they arise from some short-range order effect. The geometry of these patterns can be related to the trigonal $R\bar{3}c$ unit cell as: $[110]_c \parallel [\bar{1}\bar{2}1]_t, [101]_c \parallel [\bar{1}11]_t$, or $[011]_c \parallel [211]_t$. In any of these $<110>_c$ orientations, superlattice $½\{111\}_c$ spots, corresponding to forbidden $(\bar{1}11)_t, (101)_t$, or $(0\bar{1}1)_t$ reflections, can be explained by double diffraction in the trigonal cell.

Woodward[7] reported that this tilt system ($a^-a^-a^-$) is stabilized by highly-charged A-site cations and small tilt angles, both of which are present in $Sr_{1-3x/2}Ce_xTiO_3$.

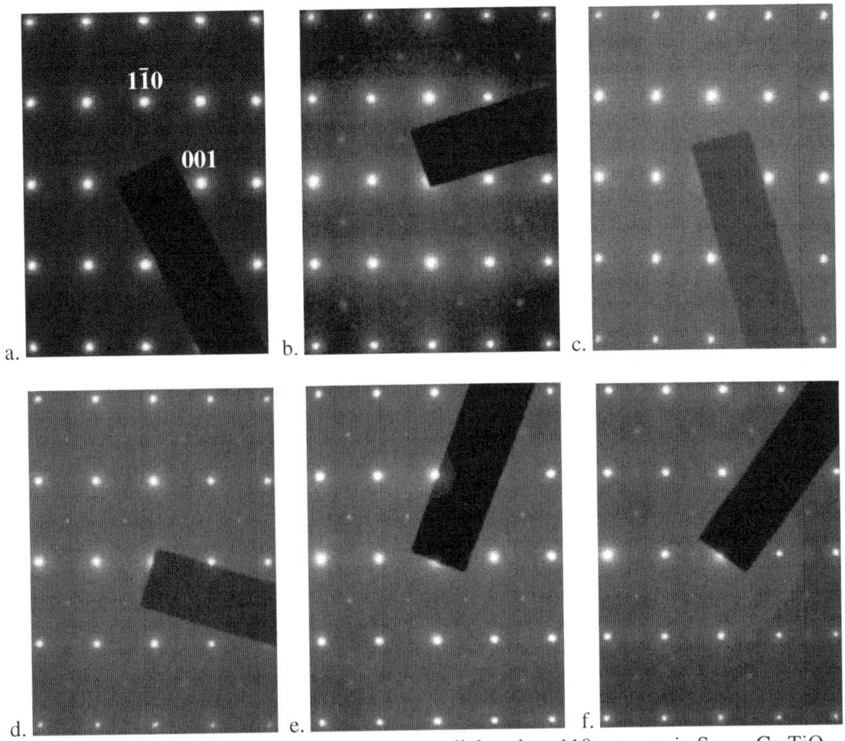

Figure 1. Electron diffraction patterns parallel to the <110>$_c$ zones in $Sr_{1-3x/2}Ce_xTiO_3$ corresponding to (a) $x = 0$, (b) $x = 0.1333$, (c) $x = 0.1667$, (d) $x = 0.1818$, (e) $x = 0.25$, and (f) $x = 0.40$.

The structure of four of these compounds was recently refined[2] in the $R\bar{3}c$ space group, and it was found that the degree of octahedral tilt increases with increasing x. Fig. 2 illustrates this result with two possible fits to the experimental data. If one assumes a data point at $x = 0$, then the trend curve obviously passes through the origin and predicts that even the smallest amount of Ce^{3+} doping will trigger octahedral tilting. On the other hand, if one allows for the possibility of a compositional cushion for very low values of x at which tilting does not occur, then the second trend curve is obtained. Even so, the revised prediction would suggest that octahedral tilting starts to occur at around $x = 0.013$. In either case, tilting seems to occur for compositions whose tolerance factors are far above the normally accepted threshold for the onset of antiphase tilting as established by Reaney *et al.*[6]

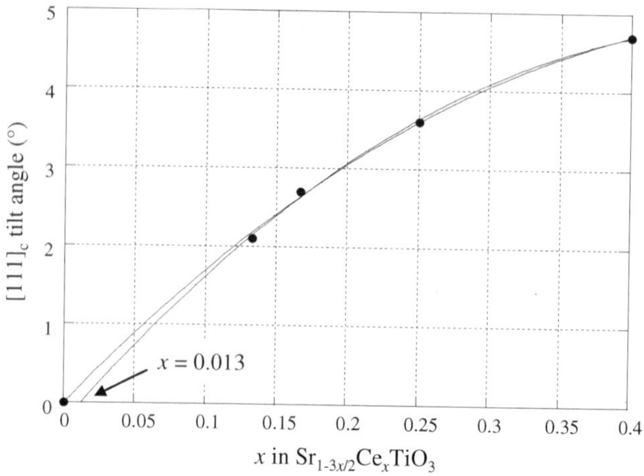

Figure 2. Degree of octahedral tilt about pseudocubic [111] as a function of x.

Based on the trigonal lattice constants published in ref. 2 for these compounds, it is possible to derive equivalent pseudocubic lattice constants according to:

$$a_c = \left(\frac{\sqrt{3}a_t^2 c_t}{12}\right)^{\frac{1}{3}} \quad (3)$$

It should also be possible to use the equation derived by Ubic[5] to calculate the pseudocubic lattice constant of these compounds:

$$a_c = 0.06741 + 0.49052(r_A + r_O) + 1.29212(r_B + r_X) \quad (4)$$

where r_A, r_B, and r_O are the ionic radii of the A, B, and O ions all in *six*-fold coordination. In fact, this equation can be re-written assuming the correct 12-fold coordination for A and two-fold coordination for O to yield an equivalent (if slightly less accurate) equation:

$$a_c = 0.05444 + 0.467016(r_A + r_X) + 1.30838(r_B + r_X) \quad (5)$$

In theory, as long as accurate values for r_A, r_B, and r_O can be obtained, then equations 3, 4, and 5 should yield equivalent results; however, the question then arises of how to calculate the radius of the A-site when it is shared by more than one species or is partially vacant, as is the case here. The concept of an "average" cation size seems at first meaningless, as each A site in this case will either be occupied by Ce^{3+}, Sr^{2+}, or a vacancy, and not by an "average" cation; however, as the strains caused by each species will be averaged over the whole structure, it is not unreasonable to expect local relaxations to allow for the stabilization of an "average" structure. If one assumes that vacancies have a size of zero and average that into the value of r_A, the inevitable result is an underestimation of a_c which increases as x increases (Figure 3). The relative error in the calculated values of a_c are $-2.74\% \leq \Delta a_c \leq -0.58\%$.

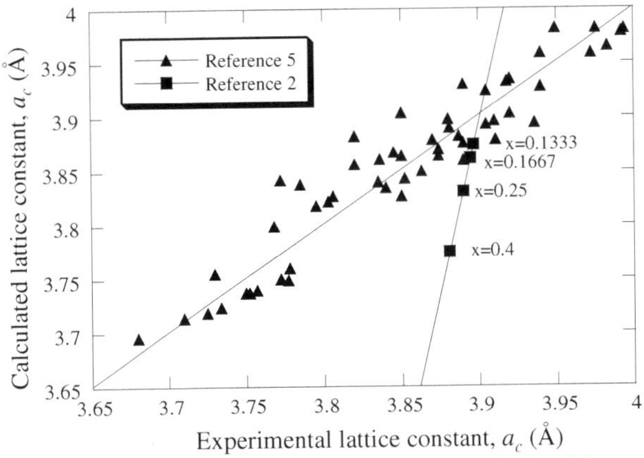

Figure 3. The pseudocubic lattice constants for $Sr_{1-3x/2}Ce_xTiO_3$ calculated from equation 4 are all underestimated.

In order to correct and understand this problem, average ionic radii can, in fact, be obtained from the refinement results in ref. 2. The values in Table 1 were obtained assuming that the substitution of Ce^{3+} on the Sr^{2+} site does not significantly affect the B-site cation.

Table I. Bond lengths and ionic radii

x	a_c (Å)	r_{AO} (Å)	r_{BO} (Å)	r_A (Å)	r_O (Å)	r_V (Å)
0.1333	3.8967	2.7557	1.9493	1.4114	1.3443	1.2110
0.1667	3.8946	2.7544	1.9488	1.4106	1.3438	1.2872
0.25	3.8900	2.7516	1.9477	1.4089	1.3427	1.3912
0.4	3.8807	2.7457	1.9447	1.4060	1.3397	1.4700

When effective values of r_O are calculated from Table 1 by subtracting r_B (presumed to be 0.605 Å)[8] from r_{BO}, it is clear that there is a very slight reduction in the average ionic radius of oxygen in this system as x increases. Oxygen ions are coordinated to two B-site cations; however, the second-nearest neighbors are four A-sites about 41% further away. An increase in the number of A-site vacancies will lower this secondary coordination of oxygen and so might be expected to cause a slight decrease in the average oxygen ionic radius. The ideal ionic radius which Shannon[8] gives is 1.35 Å, so the maximum decrease seen here is only about 0.76%.

Similarly, if one takes these derived sizes of r_O and subtracts them out from r_{AO} values, one obtains the effective size of r_A. From these values and the stoichiometry, it is possible to treat vacancies as just another ionic species and back-calculate their effective size. As Table 1 shows, vacancies have an appreciable effective size in this system. This effect might be explained by the electrostatic repulsion of the oxygen anions surrounding the vacancy.

Considering that r_{Sr} = 1.44 Å and r_{Ce} = 1.34 Å, the values of r_V are all close to the average size of the A-site cations, climbing to 98.57% of the cation size for x = 0.25. For x = 0.40, the value of r_V is not only larger than the average size of the A-site cations, but is larger even than Sr^{2+}. This result seems implausible and may in fact lend credence to the suggestion that there is a structural change at x = 0.40, as indicated by Fig. 1f.

It is now possible to derive an expression for both the effective size of A-site vacancies and the effective reduction in r_O as a function of the concentration of vacancies (Figure 4).

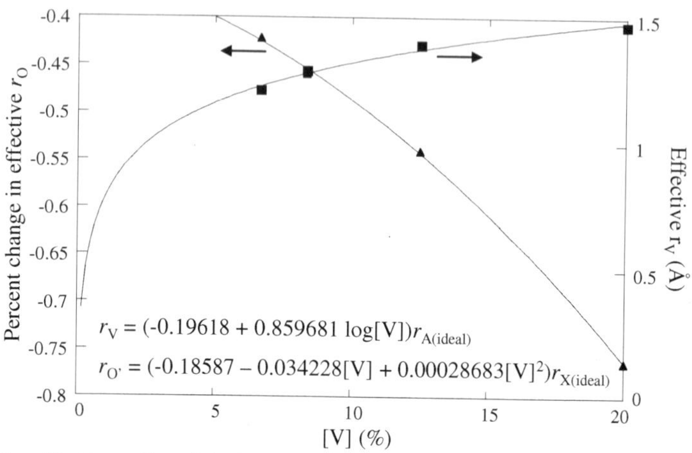

Figure 4. The effective radius of A-site vacancies (r_V) and the decrease in r_O as a function of the percentage of A-site which is vacant.

When one makes these corrections to r_V and r_O and then applies equation 5, the new predicted lattice constants have greatly improved relative errors of $-0.1257\% \leq \Delta a_c \leq 0.0103\%$. It is even possible to use the curve fits in Fig. 4 to improve the predictions for perovskites in other systems which contain A-site vacancies. For example, Ruiz *et al.*[9] reported that the cell volume of $La_{(2-x)/3}Na_xTiO_3$ seemed to increase slightly as the number of vacancies increased. The structures of the x = 0.24 and x = 0.42 compounds were refined in the orthorhombic *Ibmm* space group, and it was found that the cell volume (and so the pseudocubic lattice constant) of the x = 0.24 composition (a_c = 3.8749 Å) was slightly higher than that of the x = 0.42 composition (a_c = 3.8727 Å). Equation 4 predicts lattice constants of 3.8218 Å (Δa_c = -1.31%) for x = 0.42 and 3.7621 Å (Δa_c = -2.91%) for x = 0.24. With the corrections to r_A and r_O as derived in Fig. 4, equation 5 predicts a_c = 3.8778 Å (Δa_c = 0.1317%) for x = 0.42 and a_c = 3.8805 Å (Δa_c = 0.1445%) for x = 0.24. The unit cell expansion can be explained by the increase in the effective size of the A-site vacancies, from r_V = 0.5726 Å for x = 0.42 to r_V = 0.9875 Å for x = 0.24, which dominates the simultaneous tiny shrinkage in the average oxygen radius from 1.3501 Å to 1.3456 Å. This effective increase in r_A also explains why the tolerance factor is higher for x = 0.24 (t = 0.9888) than for x = 0.42 (t = 0.9800), thus further explaining why the degree of tilt was seen to decrease with increasing vacancy concentration.

A similar approach can be taken with the work of Arillo *et al.*[10] who studied the structure of $[(Na,K)_xLa_{(2-x)/3}](Mg_{1/2}W_{1/2})O_3$ compounds (x = 0.5, 0.335, and 0.17). They refined these structures in the monoclinic space group $P2_1/m$. Equation 4 yields errors in the pseudocubic lattice constants of $-2.96\% \leq \Delta a_c \leq -0.17\%$ for the Na analogue and $-2.55\% \leq \Delta a_c \leq 1.39\%$ for the K analogue. Again, when the corrections are made for r_A and r_O, equation 5 improves the fit errors to $0.5116\% \leq \Delta a_c \leq 0.6572\%$ for the Na family and $0.9441\% \leq \Delta a_c \leq 1.3814\%$ for K. In all these cases, the lattice constant decreases as the concentration of vacancies increases, as commonly expected.

It is possible to re-arrange equation 5 in terms of t, a_c, r_B, and r_O and then solve it for t:

$$t = \frac{a_c - 0.05444}{0.660460(r_B + r_X)} - 1.981012 \quad (6)$$

This expression should predict the effective tolerance factor of a perovskite given only the cubic/pseudocubic lattice constant, the ionic radius of the B-site cation (Ti^{4+} in this case), and the effective ionic radius of the X anion (O^{2-} in this case). It removes the necessity of calculating effective or average A-site ionic radii for compositions with mixed occupancies or A-site vacancies.

In the case of $Sr_{1-3x/2}Ce_xTiO_3$, equation 2 would yield tolerance factors of 1.0091, 0.9696, 0.9597, 0.935, and 0.8905 for x = 0, 0.1333, 0.1667, 0.25, and 0.40, respectively, assuming a vacancy size of zero. In all but the case of x = 0 (where there are no vacancies), the predicted structure would exhibit antiphase octahedral tilting. In the case of $x \geq 0.1667$ the structures predicted would also exhibit in-phase tilting; however, it has already been established[2] that no in-phase tilting exists in these perovskites. If instead one simply ignores vacancies and takes an average of the radii of the actual A-site species as r_A, then the values are all much higher ($0.9910 \leq t \leq 1.0091$). Another possibility would be to calculate ionic radii from the refined structural models and plug them into equation 2. In this case the tolerance factors are all very close to unity ($0.9984 \leq t \leq 1$). This phenomenon is in fact generally true for tilted perovskites. A tolerance factor of 1 is energetically favorable, and so structures which would otherwise have higher or lower tolerance factors deform in such a way as to bring them closer to unity. Such values of tolerance factor are interesting from a structural point of view but are useless in a predictive sense, as they require the structural refinement of an already-synthesized material in order to calculate. A more useful predictive form of the tolerance factor is given in equation 6. With this equation, it can be shown that the tolerance factors of $Sr_{1-3x/2}Ce_xTiO_3$ are $0.9980 \leq t \leq 1.0034$ for $0 \leq x \leq 0.40$.

The model of Reaney *et al.*[6] for niobates and tantalates would predict an untilted structure for such high values of t. On the other hand, many researchers have reported aluminates,[11] cuprates,[12] nickelates,[13] and ferrites[14] with $t \sim 1$ and tilted structures. For example, $LaAlO_3$ has been reported in the $R\bar{3}c$ system[11] with octahedra tilted 5.0° about the $[111]_c$, yet it has a t = 1.0166 (or 1.0198 according to equation 6). According to Woodward,[7] the rhombohedral $a^-a^-a^-$ tilt system is stabilized by highly charged A-site cations (which were not a feature of the materials in ref. 6) and small tilt angles. The $\langle 111 \rangle_c$ tilt in $Sr_{1-3x/2}Ce_xTiO_3$ ($0 \leq x \leq 0.40$) materials is only $0 \leq \phi \leq 4.7°$, which is even smaller than that reported in $LaAlO_3$. At higher tilt angles, the orthorhombic $a^+b^-b^-$ tilt system should be preferred, as in the case of $La(Zn_{1/2}Ti_{1/2})O_3$, which has an in-phase tilt angle of 8.3°.[15]

CONCLUSIONS

Four compositions in the $Sr_{2-3x/2}Ce_xTiO_3$ homologous series corresponding to x = 0.1333, 0.1667, 0.1818, 0.25, and 0.4 contain a non-cubic perovskite superlattice caused by antiphase octahedral tilting. The degree of tilt increases with increasing x, and the probable space group is $R\bar{3}c$. A-site vacancies in this system have an effective size greater than zero, presumably due to mutual repulsion of oxygen ions adjacent to the vacancies. Such vacancies also cause a very slight reduction in the radius of oxygen anions as a result of the lowering of their secondary coordination. This same phenomenology has been successfully applied to other systems to explain errors in the predicted pseudocubic lattice constants.

REFERENCES

[1]G. Subodh, J. James, M.T. Sebastian, R. Paniago, A. Dias, and R.L. Moreira, Structure and Microwave Dielectric Properties of $Sr_{2+n}Ce_2Ti_{5+n}O_{15+3n}$ ($n \leq 10$) Homologous Series, *Chem. Mater.*, **19**, 4077-82 (2007).

[2]R. Ubic, G. Subodh, M.T. Sebastian, D. Gout, and T. Proffen, Structure of Compounds in the $Sr_{1-3x/2}Ce_xTiO_3$ Homologous Series, *Chem.Mater.*, **20**, 3127-33 (2008).

[3]C.E. Bamberger, T.J. Haverlock, and O.C. Kopp, Synthesis and Characterization of $Sr_2Ce_2Ti_5O_{16}$ in the System SrO-CeO_2-TiO_2, *J. Am. Ceram. Soc.*, **77**, 1659-61 (1994).

[4]R.L. Moreira, R.P.S.M. Lobo, G. Subodh, M.T. Sebastian, F.M. Matinaga, and A. Dias, Optical Phonon Modes and Dielectric Behavior of $Sr_{1-3x/2}Ce_xTiO_3$ Microwave Ceramics, *Chem. Mater.*, **19**, 6548-54 (2007).

[5]R. Ubic, Revised Method for the Prediction of Lattice Constants in Cubic and Pseudocubic Perovskites, *J. Am. Ceram. Soc.*, **90**, 3326-30 (2007).

[6]I.M. Reaney, E.L. Colla, and N. Setter, Dielectric and Structural Characteristics of Ba- and Sr-Based Complex Perovskites as a Function of Tolerance Factor, *Jpn. J. Appl. Phys., Part 1*, **33**, 3984-90 (1994).

[7]P.M. Woodward, Octahedral Tilting in Perovskites. II. Structure Stabilizing Forces, *Acta Cryst.*, **B53**, 44-66 (1997).

[8]R.D. Shannon, Revised Effective Ionic Radii and Systematic Studies of Interatomic Distances in Halides and Chalcogenides, *Acta Cryst.*, **A32**, 751-767 (1976).

[9]A.I. Ruiz, M.L. López, C. Pico, and M.L. Veiga, New $La_{2/3}TiO_3$ Derivatives: Structure and Impedance Spectroscopy, *J. Solid State Chem.*, **163**, 472-478 (2002).

[10]M.A. Arillo, J. Gómez, M.L. López, C. Pico, and M.L. Veiga, Structural and Electrical Characterization of New Materials with Perovskite Structure, *Solid State Ionics*, **95**, 241-248 (1997).

[11]C.J. Howard, B.J. Kennedy, and B.C. Chakoumakos, Neutron Powder Diffraction Study of Rhombohedral Rare-Earth Aluminates and the Rhombohedral to Cubic Phase Transition, *J. Phys.: Condens. Matter*, **12**, 349-365 (2000).

[12]G. Demazeau, C. Parent, M. Pouchard, and P. Hagenmuller, Sur Deux Nouvelles Phases Oxygenees du Cuivre Trivalent. $LaCuO_3$ et $La_2Li_{0,50}Cu_{0,50}O_4$, *Mater. Res. Bull.*, **7**, 913-920 (1972).

[13]J.L. García-Muñoz, J. Rodriguez-Carvajal, P. Lacorre, and J.B. Torrance, Neutron-Diffraction Study of $RNiO_3$ (R= La, Pr, Nd, Sm): Electronically Induced Structural Changes Across the Metal-Insulator Transition, *Phys. Rev. B: Condens. Matter*, **46**, 4414-25 (1992).

[14]P.D. Battle, T.C. Gibb, and P. Lightfoot, The Structural Consequences of Charge Disproportionation in Mixed-Valence Iron Oxides. I. The Crystal Structure of $Sr_2LaFe_3O_{8.94}$ at Room Temperature and 50K, *J. Solid State Chem.*, **84**, 271-279 (1990).
[15]R. Ubic, Y. Hu, and I. Abrahams, Neutron and Electron Diffraction Studies of $La(Zn_{1/2}Ti_{1/2})O_3$ Perovskite, *Acta Cryst.*, **B62**, 521-529 (2006).

ACKNOWLEDGEMENTS
This work has been supported by the National Science Foundation through the Major Research Instrumentation Program, Award Number 0521315, and the US Agency for International Development, award number PGA-P280420. The authors are also indebted to Steve Letourneau of Boise State University for TEM sample preparation.

EFFECT OF DOPANTS AND PROCESSING ON THE MICROSTRUCTURE AND DIELECTRIC PROPERTIES OF $CaCu_3Ti_4O_{12}$ (CCTO)

B. Bender and M. Pan
Naval Research Laboratory, Washington, DC, USA

ABSTRACT

Research has been conducted on the giant dielectric constant oxide CCTO and it has shown that the dielectric material has a giant permittivity (as large as 80,000) that is stable over a range of temperatures and frequencies. However, the typical CCTO ceramic usually has a dielectric loss of 0.1 or higher which needs to be reduced if this dielectric oxide is going to be used commercially. This research has shown that slow cooling in oxygen and annealing at 1000°C can improve the dielectric loss properties. Doping with PbO also improved loss properties and lead to a two-fold increase in breakdown voltage. Co-doping with hafnia and calcia followed by annealing avoided sacrificing permittivity for loss leading to a $CaCu_3Ti_4O_{12}$ ceramic with a giant dielectric constant of 69,000 and a tan δ of 0.027.

INTRODUCTION

The Navy in a drive to make their ships and combat vehicles more efficient and more effective has instituted research programs to develop the all-electric ship. The all-electric ship will be designed to have the propulsion, auxiliary and weapons systems drawing from the same energy source. To achieve this goal the Navy has conducted extensive research into power electronics. Considerable progress has been made in developing state-of-the-art power converters but the size of the filter capacitors is still a limiting factor in decreasing the footprint of the power converter. To decrease the size of the capacitors their permittivity must be enhanced. The ideal filter capacitor should also be stable over a range of temperatures in regard to frequency and voltage. With typical commercial dielectrics such as $BaTiO_3$ one usually has to sacrifice permittivity for stability. However, recent research on $CaCu_3Ti_4O_{12}$ has shown that this dielectric oxide has the potential to be an ideal capacitor material. Subramanian *et al.* [1] were the first to measure CCTO's dielectric properties and they found that polycrystalline $CaCu_3Ti_4O_{12}$ possesses a dielectric constant of 12,000 (room temperature- 1 kHz) that exhibits little temperature dependence from zero to 200°C. Permittivity measurements on single crystal CCTO showed a giant dielectric constant as high as 80,000 [2]. Also the material can be engineered into an internal barrier layer capacitive-like (IBLC) dielectric via one-step processing in air at modest sintering temperatures of 1050 to 1100°C [3].

However, for this material to be used commercially its dielectric loss properties have to be improved. Dielectric loss values as low as 0.05 (1 kHz) have been reported [4,5]. Unfortunately, the loss of these CCTO ceramics is very sensitive to temperature as tan δ-temperature curves start to warp up at temperatures as low as 40°C leading to losses that surpass 0.10 before 60°C is reached [4, 6-8]. Before the dielectric loss properties can be improved the true nature of the giant permittivity of $CaCu_3Ti_4O_{12}$ has to be discerned. Many different explanations have been put forth [9] ranging from electrode depletion effects [10] to relaxor-like dynamical slowing of dipolar fluctuations in nanosize domains [2]. Most researchers believe that

CCTO's giant permittivity is extrinsic in nature and is the result of the formation of internal capacitive barrier layers [1,3,11]. Researchers believe that insulating surfaces form during processing at the grain boundaries of semiconducting grains. This creates an electronically heterogeneous material comparable to internal barrier layer capacitors (IBLCs). Chung *et al.* [12] showed conclusively that a large electrostatic potential barrier exists at the grain boundaries in CCTO with n-type semiconducting grains separated by an insulative grain boundary region. However, the exact mechanisms responsible for the semiconducting nature of the grains and the chemistry and defect nature of the grain boundary region are still unclear. The electrical measurements of CCTO by Adams *et al.* [13] and Zang *et al.* [14] show that the electrostatic potential barrier can be best described using a double Schottky barrier (DSB) model. DSB's play an important role in the properties of varistors and PTCR's (positive temperature coefficient resistors). A myriad of research has been done showing the importance of processing on optimizing the properties of varistors and PTCRs [15-17]. Clarke [15] in his review of varistors brings out the importance of oxygen, cooling schedules, annealing, and microstructural engineering in the optimization of ZnO varistors. Buchanan [16] discusses the importance of doping on the non-linear voltage properties of varistors as up to eight dopants are used to optimize the performance of commercial ZnO varistors. This paper reports the effect of slow cooling in oxygen, annealing in air, and the effect of dopants (lead oxide, hafnia, and calcia) on the dielectric properties of various $CaCu_3Ti_4O_{12}$ ceramics.

EXPERIMENTAL PROCEDURE

$CaCu_3Ti_4O_{12}$ was prepared using ceramic solid state reaction processing techniques. Stoichiometric amounts of $CaCO_3$ (99.98%), CuO (99.5%) and TiO_2 (99.5%) were mixed by blending the precursor powders into a purified water solution containing a dispersant (Tamol 901) and a surfactant (Triton CF-10). The resultant slurries were then attrition-milled for 1 h and dried at 90°C. The standard processed powder, ccto05, was calcined at 900°C for 4 h and then 945°C for 4 h. After the final calcination the ccto05 powders were attrition-milled for 1 h to produce finer powders. The PbO-doped powder, pcto, was fabricated by mixing 0.37 w/o PbO (99.9%) with the calcined ccto05 powder. The calcia-doped powder, cacto, and hafnia-doped powder (hcto), were made by mixing 0.3 w/o $CaCO_3$ (99.98%) or 0.2% HfO_2 (99.5%) with the calcined ccto05 powder. Co-doped powders (cahcto) were made by blending equal amounts of hcto and cacto powders. A 2% PVA binder solution was mixed with the powders and they were sieved to eliminate large agglomerates. The dried powder was uniaxially pressed into discs typically 13 mm in diameter and 1 mm in thickness. The discs were then placed on platinum foil and sintered in air for the standard time of 16 h. The slow-cooled (sccto) samples were cooled from 1100°C to 750°C at 30°C/h in flowing oxygen. The annealed samples were heat-treated after sintering in air for 12 h at 1000°C.

Material characterization was done on the discs and powders after each processing step. XRD was used to monitor phase evolution for the various mixed powders and resultant discs. Microstructural characterization was done on the fracture surfaces using scanning electron microscopy (SEM). To measure the dielectric properties, sintered pellets were ground and polished to achieve flat and parallel surfaces onto which palladium-gold electrodes were sputtered. The capacitance and dielectric loss of each sample were measured as a function of temperature (-55 to 120°C) and frequency (100 Hz to 100 KHz) using an integrated, computer-controlled system in combination with a Hewlett-Packard 4284A LCR meter. Electrical breakdown was measured on samples typically 1 mm in thickness with gold electrodes at an applied rate of voltage of 500 volts per second.

RESULTS AND DISCUSSION

The Effect of Slow Cooling in Oxygen on Electrical Properties of CCTO

Slow cooling in oxygen did affect the electrical properties of $CaCu_3Ti_4O_{12}$. Permittivity (see Fig. 1- 20°C) dropped by 27% from 40,000 to 29,000. The dielectric loss increased by 22% going from 0.018 to 0.022 (see Table I). However, slow oxygen cooling dramatically affected the temperature stability of the dissipation factor as with a 60°C rise in temperature from room temperature the dielectric loss increased 3-fold for as-processed CCTO versus only 50% (0.033) for the slow-cooled sccto. Also the measured breakdown voltage increased (Table I) increased by 30% when the sample was slow-cooled in oxygen. Last of all, slow-cooling affected the grain boundaries leading to a definite increase in the amount of intergranular fracture (see Fig. 2).

Table I Dielectric and Microstructural Properties of various ccto05 $CaCu_3Ti_4O_{12}$ Ceramics

Material	K (1 kHz)	Tan δ (1 kHz)	Breakdown Volt. (V/cm)	Relative Density	Median Grain Size (μm)
As-processed	40039	0.018	1120	96.0%	31
Slow-cooled	29263	0.022	1450	96.0%	31
32 h sintered	47286	0.068	N.M.	N.M.	46
32 h + anneal	40771	0.027	N.M.	N.M.	46
PbO-doped	15081	0.025	3120	94.3%	28
HfO2-doped	30630	0.016	N.M.	94.4%	30
CaO-doped	87372	0.080	N.M.	93.5%	50
Hf/Ca-doped	80331	0.071	N.M.	95.4%	46
Hf/Ca+ anneal	69042	0.027	N.M.	95.4%	46

It is believed that an increase in oxygen adsorption is occurring at the grain boundaries during slow cooling which is contributing to the measured changes found in the sccto material. It is well known that chemisorption of oxygen ions can dramatically affect the DSB properties in titanate and ZnO-based varistors [15,18]. Bueno *et al.* [19] believe that in oxide-based varistors that the grain boundary becomes more oxidized when treated in an oxygen-rich atmosphere which increases the potential of the Schottky barriers. This rise can cause the effective resistivity of the grain boundaries to increase which would lead to an increase in the temperature stability of tan δ [20]. Indirect evidence that the grain boundary resistivity is being affected can be found by examining the effect of frequency on the dielectric loss properties. In an IBLC the approximation for the loss tangent is [21]

$$\tan\delta = 1/(\omega R_{gb}C) + \omega R_g C \quad (1)$$

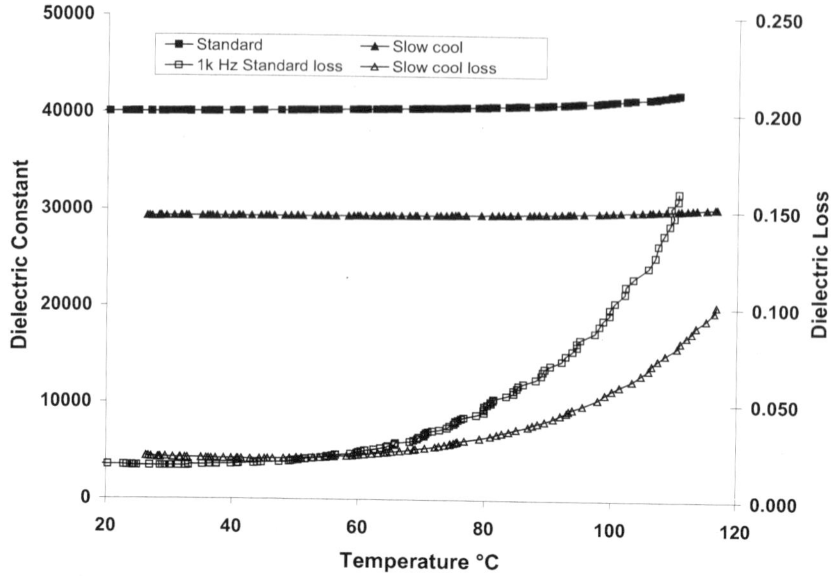

Fig. 1 Temperature dependence of the dielectric constant and loss for as-processed ccto05 and ccto05 slow cooled in flowing oxygen

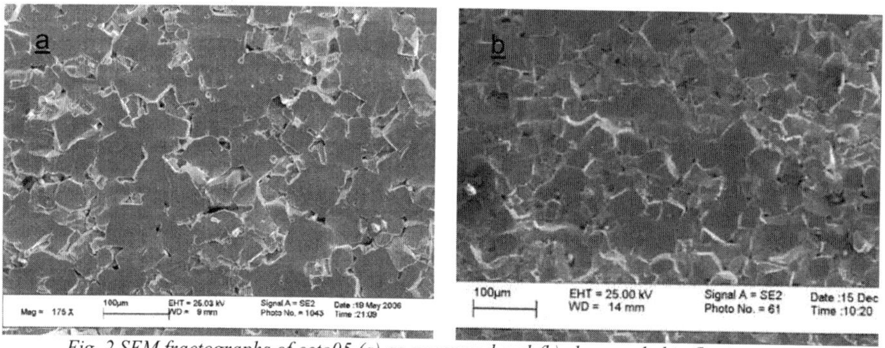

Fig. 2 SEM fractographs of ccto05 (a) as-processed and (b) slow-cooled in flowing oxygen

where ω is the angular frequency, R_{gb} is the resistance of the boundary layer, and R_g is the resistance of the semiconducting grains. This means that the low frequency loss is dominated by R_{gb} while higher frequency loss is dominated by R_g. Detailed examination of the dielectric spectra reveals that the changes in loss of sccto as compared to ccto05 are greater at 100 Hz. This implies that slow cooling in oxygen is indeed affecting the resistivity of the grain boundaries. Also Chung *et al.* [22] have done impedance spectroscopy (IS) research on doped-

CCTO and have shown a direct correlation between changes in R_{gb} and E_b. This corresponds well with our measured electrical breakdown data showing a substantial increase in E_b for the slow-cooled sample.

Effect of 1000°C Annealing on the Dielectric Properties of $CaCu_3Ti_4O_{12}$

Annealing in air at 1000°C for 12 h did affect the electrical properties of $CaCu_3Ti_4O_{12}$. Dielectric constant (1 kHz at 20°C) dropped by 14% while tan δ dropped 60% to 0.027 (Table I). However, annealing dramatically affected the temperature stability of tan δ as shown in Fig. 3. Annealing did not affect the grain size or amount of porosity but did affect the grain boundaries leading to a sizeable increase in the amount of intergranular fractured when annealed at 1000°C.

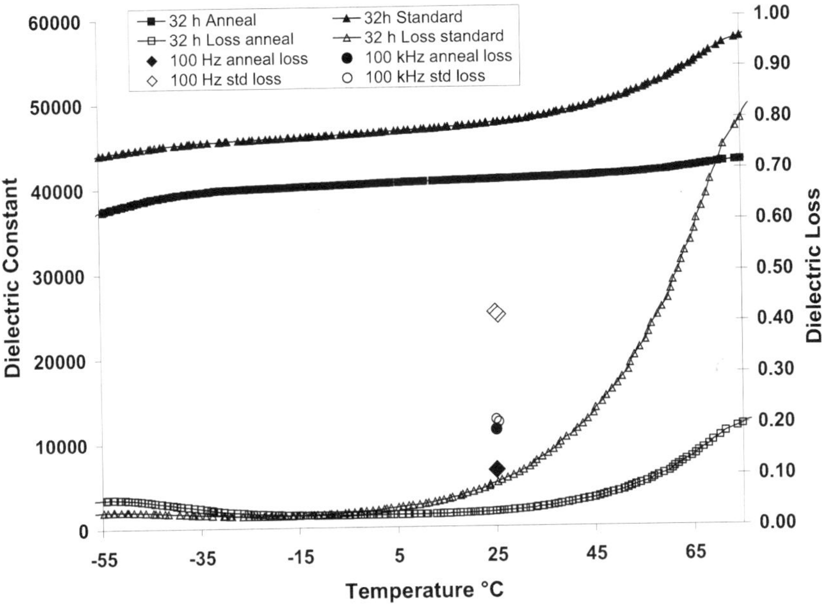

Fig. 3 Temperature dependence of dielectric constant and loss (1 kHz) for ccto05 sintered for 32 hours and sintered for 32 hours and annealed for 1000°C for 12 hours

Similar to the slow-cooled sccto material it is believed that oxygen adsorption at the grain boundaries is occurring which affects the chemistry and the resistivity of the grain boundaries and the resultant properties of the CCTO. Annealing in different atmospheres does affect the dielectric properties of $CaCu_3Ti_4O_{12}$. Bender *et al.* [4] showed that dramatic increases in permittivity and loss occurred when CCTO was annealed in flowing argon. Adams *et al.* [23] showed similar results for nitrogen but they demonstrated that the dielectric properties could be totally reversed when reannealed in flowing oxygen but only at a high enough temperature (1000°C). They also showed via IS that reannealing in oxygen led to higher grain boundary resistivities. Therefore it is possible that annealing at 1000°C in air is resulting in oxygen adsorption at the grain boundaries which would make the boundaries more insulative. From equation one if the annealing did affect the grain boundary resistivity it would result in greater changes in the effect of dielectric loss at lower frequencies than at higher frequencies. Fig. 3 dramatically shows this as the large circles show the small difference in loss values at 100 kHz versus the substantial difference in loss values at 100 Hz as represented by the large squares. The higher grain boundary resistivity would result in the improvement in the dielectric loss properties of the $CaCu_3Ti_4O_{12}$. The long anneal times could also lead to an effective thicker boundary layer which would improve loss properties too. A thicker boundary layer would also explain the reduction in permittivity. The dielectric constant in an IBLC or material with a Schottky barrier layer can be represented as [14]

$$\varepsilon' = \varepsilon_B \, (d/t_B) \quad (2)$$

where ε_B is the permittivity of the boundary or barrier layer, d can be approximated by the grain size, and t_B is the effective thickness of the boundary or barrier layer. Since the grain size of the annealed sample remains the same during annealing and if the thickness of the boundary layer increased then the permittivity would drop which is what is observed.

Effect of Doping with PbO on the Dielectric Properties of $CaCu_3Ti_4O_{12}$

Doping CCTO with 0.37% PbO affected the resultant microstructure and dielectric properties of the standard ccto05 powders. The median grain sizes are similar (31 vs. 28 microns for pcto- see Table I) while the relative density of the pcto sample dropped from 96% to 94.3%. As seen in comparing Fig. 4a to Fig. 2b, the PbO-doped sample fractures in a more intergranular-fashion. Also in Fig. 4b grain boundary phases are more prominent which may explain the increase in intergranular fracture. EDS analysis indicates that these grain boundary phases are PbO-rich. Dielectric measurements (see Fig. 5) show that PbO-doping dramatically affects the permittivity and dielectric loss properties of $CaCu_3Ti_4O_{12}$. The PbO-doped sample dielectric constant drops by 60% from 40,000 to 15,000. However, the dielectric loss temperature stability properties are greatly improved as the loss at 106°C is only 0.059 which is 50% of the value of the standard ccto05 material.

The implication from the above data is that doping with PbO is increasing the effective grain boundary thickness. Since the grain size is relatively unchanged, equation 2 implies that the effective barrier layer thickness would have to be significantly increased. A thicker grain boundary can lead to improved tan δ properties. This has been shown to be true in the case of doping CCTO with $CaTiO_3$. Yan *et al.* [21] showed that at a certain level of doping that the effective thickness of the grain boundaries reached a point that inhibited electron tunneling effects, and at that level, the resistance of the grain boundaries increased exponentially. That this may be the case for the pcto samples is exemplified by the fact the breakdown voltage for pcto increased more than 2-fold from 1450 to 3120 V/cm. The thicker effective barrier layers may be the result of the presence of the more prevalent Pb-rich phases observed at the grain boundaries of the pcto ceramics (see Fig. 4b). The resultant pcto ceramic exhibits excellent X7R capacitor properties as it has a giant dielectric constant of 15,000 at 20°C (1 kHz) whose permittivity varies only ± 7.5% between -60 and 125°C and only ± 6% with frequency from 100 Hz to 100 kHz and whose tan δ varies only from 0.020 to 0.076 over the same temperature range.

Fig. 4 SEM fractographs of PbO-doped $CaCu_3Ti_4O_{12}$ showing (a) large amounts of intergranular fracture and (b) the presence of a Pb-rich grain boundary phase present at some of the grain boundaries

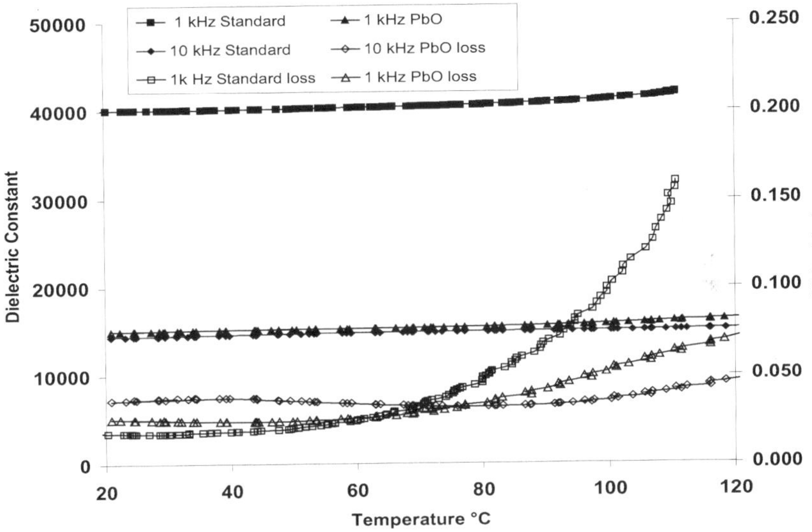

Fig. 5 Temperature dependence of dielectric constant and loss of standard cct05 and ccto05 doped with PbO

Effect of Co-doping with CaO and HfO_2 on the Dielectric Properties of $CaCu_3Ti_4O_{12}$

Previous research by Bender *et al.* [24] had shown that doping CCTO with either calcia or hafnia led to $CaCu_3Ti_4O_{12}$ ceramics with differing dielectric properties as shown in Fig. 6. Doping with hafnia led to a CCTO with lower permittivity and a lower tan δ and an increase in dielectric loss thermal stability. Doping with CaO led to a significant increase in grain size (by 60%- see Fig. 7a & b) combined with a considerable increase in permittivity and loss. Co-doping CCTO was undertaken in an attempt to increase permittivity from the standard ccto05 without sacrificing dielectric loss. Dielectric measurements indicate that co-doping was a partial success. The dielectric constant of the composite dropped only 10% from that of cacto material and was 70% greater than the standard ccto05 ceramic (see Table I). However, the tan δ increased from 0.018 to 0.071 and its loss temperature stability was significantly less than for the undoped CCTO. Annealing at 1000°C for 12 h improved the properties of the composite dielectric even more. The dielectric loss dropped to 0.027 and the temperature stability of the loss improved noticeably while undergoing only a 20% loss in permittivity to 69,000 (see Fig. 6). Microstructural observations showed (Fig. 7c & d) that the co-doped ceramic was very similar in microstructure to that of the cacto ceramic with the median grain size dropping from 50 to 46 microns for the co-doped CCTO.

Analysis of the microstructural and the dielectric properties indicates that the co-doped samples partially retain the favorable benefits of both dopants. The 8% drop in median grain size correlates well with the 10% decrease in dielectric constant for cahcto. Detailed examination of the dielectric spectra indicate that the grain boundary resistivity was affected by co-doping and parallels the increase in grain boundary resistivity observed for hafnia-doped samples [24]. Annealing led to further increases in the grain boundary resistivity as evidenced by the drop in tan δ and the improvement in the thermal stability of the loss factor. As explained earlier it is possible that chemisorption of oxygen ions is occurring at the grain boundary which would lead to a more insulative boundary. It is also possible the long annealing times are reducing the heterogeneity of the grain boundary segregants which would lead to a thicker more robust more uniform boundary layer which would also increase the resistivity of the grain boundary while at the same time reduce the permittivity of the dielectric which is what was measured (see Fig. 6 and Table I). The resultant CCTO ceramic has excellent capacitive properties exhibiting a giant dielectric constant at room temperature (1 kHz) of 69,000 with a loss of only 0.027 and good thermal stability in regards to both permittivity and dielectric loss.

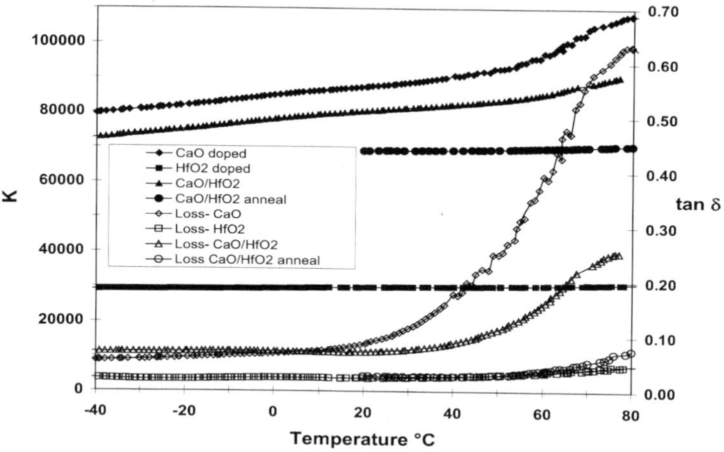

Fig. 6 Temperature dependence of dielectric constant and loss for ccto5 doped with calcia, hafnia, calcia and hafnia, and calcia and hafnia plus being annealed at 1000°C for 12 hours

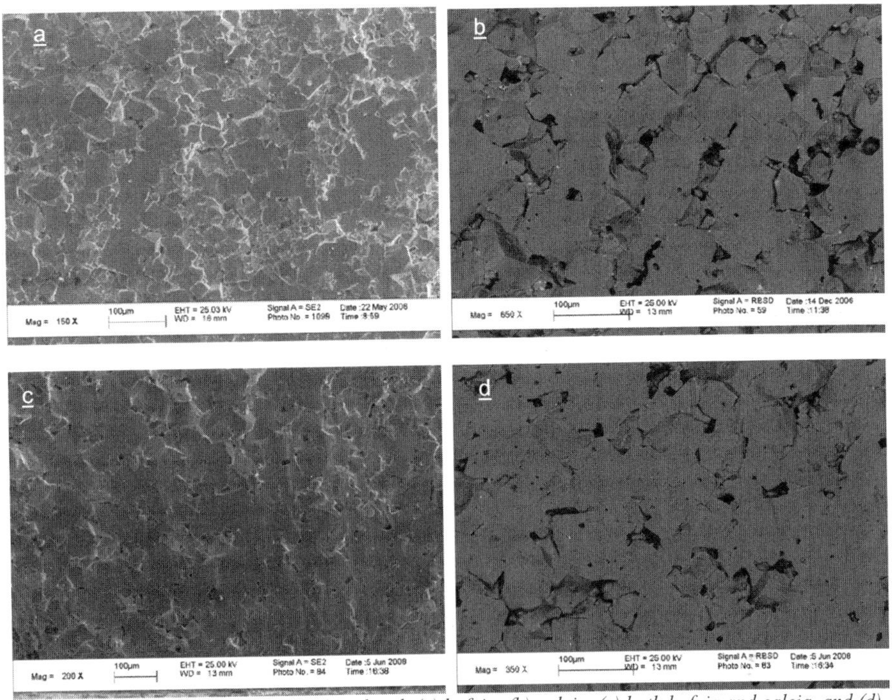

Fig. 7 SEM fractographs of ccto05 doped with (a) hafnia, (b) calcia, (c) both hafnia and calcia, and (d) co-doped with hafnia and calcia and annealed at 1000°C for 12h

CONCLUSIONS

Research has shown that slow cooling in flowing oxygen and annealing at long times at 1000°C has improved dielectric loss properties at the expense of lower permittivity. Evidence points toward the role of oxygen adsorption at the grain boundaries during these heat treatments which increases the resistivity of the grain boundary. This leads to higher breakdown voltage and improved thermal stability of tan δ. Doping $CaCu_3Ti_4O_{12}$ with PbO leads to a dielectric with more effective insulative boundary barriers resulting in a capacitor material that meets X7R specifications with a giant dielectric constant of 15,000. Co-doping with hafnia and calcia led to a CCTO that partially retained the favorable properties of the individual dopants. Annealing in air led to an improved dielectric material which was able to retain high permittivity without sacrificing loss properties resulting in a CCTO with a dielectric constant of 69,000 and a loss of only 0.027. Further IS work and transmission electron microscopy would be beneficial in developing a better understanding of the role of the dopants and oxygen on the chemistry and the electrical properties of the boundary and barrier layers. This could lead to microstructural and processing engineering of CCTO that would result in the full optimization of the giant dielectric constant properties of $CaCu_3Ti_4O_{12}$.

REFERENCES

[1] M. Subramanian, D. Li, N. Duan, B. Reisner, and A. Sleight, High Dielectric Constant in $ACu_3Ti_4O_{12}$ and $ACu_3Fe_3O_{12}$ Phases, *J. of Solid State Chem.*, Vol 151, 2000, p 323-25

[2] C. Homes, T. Vogt, S. Shapiro, S. Wakitomo, A.Ramirez, Optical Response of High-Dielectric-Constant Perovskite-Related Oxide, *Science*, Vol 293, 2001, p 673-76

[3] T. Adams, D. Sinclair, and A. West, Giant Barrier Layer Capacitance Effects in $CaCu_3Ti_4O_{12}$ Ceramics, *Adv. Mater.*, Vol 14, 2002, p 1321-23

[4] B. Bender and M. Pan, The Effect of Processing on the Giant Dielectric Properties of $CaCu_3Ti_4O_{12}$, *Mat. Sci. Eng. B.*, Vol 117, 2005, p 339-47

[5] E. Patterson, S. Kwon, C. Huang, and D. Cann, Effects of ZrO_2 Additions on the Dielectric Properties of $CaCu_3Ti_4O_{12}$, *Appl. Phys. Lett.*, Vol 87, 2005, p 182911-1-3

[6] T. Fang, L. Mei, H. Ho, Effects of Cu Stoichiometry on the Microstructures, Barrier-Layer Structures, Electrical Conduction, Dielectric Responses, and Stability of $CaCu_3Ti_4O_{12}$, *Acta Mat.*, Vol 54, 2006, p 2867-75

[7] R. Grubbs, E. Venturini, P. Clem, J. Richardson, B. Tuttle, and G. Samara, Dielectric and Magnetic Properties of Fe- and Nb-doped $CaCu_3Ti_4O_{12}$, *Phys. Rev. B*, Vol 72, 2005, p 104111-1

[8] T. Fang and H. Shiau, Mechanism for Developing the Boundary Barrier Layers of $CaCu_3Ti_4O_{12}$, *J. Am. Ceram. Soc.*, Vol 87, 2004, p 2072-79

[9] W. Li, R. Schwart, A. Chen, and J. Zhu, Dielectric Response of Sr-Doped $CaCu_3Ti_4O_{12}$ Ceramics, *Appl. Phys. Lett.*, Vol 90, 2007, p 122901-1-3

[10] P. Lunkenheimer, R. Fichtl, S. Ebbinghau, and A. Loidl, Nonintrinsic Origin of the Colossal Dielectric Constant in $CaCu_3Ti_4O_{12}$, *Phys. Rev. B*, Vol 70, 2004, p 172102

[11] L. He, J. Neaton, M. Cohen, and D. Vanderbilt, First-Principles Study of the Structure and Lattice Dielectric Response of $CaCu_3Ti_4O_{12}$, *Phys. Rev.*, Vol 65, 2002, p 214112-1-11

[12] S. Chung, I. Kim, and S. Kang, Strong Nonlinear Current-Voltage Behaviour in Perovskite-Derivative Calcium Copper Titanate, *Nature Materials*, Vol 3, 2004, p 774-78

[13] T. Adams, D. Sinclair, A. West, Characterization of Grain Boundary Impedances in Fine- and Coarse-Grained $CaCu_3Ti_4O_{12}$ Ceramics, *Phy. Rev. B*, Vol 73, 2006, p. 094124-1

[14] G. Zang, J. Zhang, P. Zheng, J. Wang, and C. Wang, Grain Boundary Effect on the Dielectric Properties of $CaCu_3Ti_4O_{12}$ Ceramics, *J. Phys. D. Appl. Phys.*, Vol **38**, 2005, p 1824-27

[15] D. Clarke, Varistor Ceramics, *J. Am. Ceram. Soc.*, Vol 82, 1999, p 485-502

[16] R.Buchanan, *Ceramic Materials for Electronics*, Marcel Dekker, 2004, p 377-431

[17] G. Lewis, C. Catlow, and R. Casselton, PTCR Effect in $BaTiO_3$, *J. Am. Ceram. Soc.*, Vol 68, 1985, p 555-58

[18] J. Li, S. Li, F. Liu, M. Alim, G. Chen, The Origin of Varistor Property of $SrTiO_3$-Based Ceramics, *J. Mat. Sci.: Mat. Electr.*, Vol 14, 2003, p 483-486

[19] P. Bueno, E. Leite, M. Oliveira, M. Orlandi, and E. Longo, Role of Oxygen at the Grain Boundary of Metal Oxide Varistors: A Potential Barrier Formation Mechanism, *Appl. Phys. Lett.*, Vol 79, 2001, p 48-50

[20] D. Cann, E. Patterson, S. Kwon, C. Huang, S. Martin, High-K $CaCu_3Ti_4O_{12}$ Dielectrics, *Extended Abstracts of the 12th US-Japan Seminar on Dielectric and Piezoelectric Ceramics*, 2003, p 309-312

[21] Y. Yan, L. Jin, L. Feng, and G. Cao, Decrease of Dielectric Loss in Giant Dielectric Constant $CaCu_3Ti_4O_{12}$ Ceramics by Adding $CaTiO_3$, *Mat. Sci. Eng. B*, Vol 130, 2006, p 146-50

[22] S. Chung, S. Lee, J. Choi, and S. Choi, Initial Cation Stoichiometry and Current-Voltage Behavior in Sc-Doped Calcium Copper Titanate, *Appl. Phys. Lett.*, Vol 89, 2006, p 191907-1-3

[23] T. Adams, D. Sinclair, and A. West, Influence of Processing Conditions on the Electrical Properties of $CaCu_3Ti_4O_{12}$ Ceramics, *J. Am. Ceram. Soc.*, Vol 89, 2006, p 3129-3135

[24] B. Bender and M. Pan, The Effect of Sintering Conditions and Dopants on the Dielectric Loss of the Giant Dielectric Constant Perovskite $CaCu_3Ti_4O_{12}$, *Ceram. Eng. Sci. Proc.*, Vol 28, 2007, p 87-98

Author Index